P9-CFT-742

NOT
BY
CHANCE!

Shattering
the Modern Theory
of Evolution

---

by
Lee M. Spetner, PhD

The Judaica Press, Inc.

ISBN 1-880582-24-4

THE JUDAICA PRESS, INC.
123 Ditmas Avenue Brooklyn, New York 11218
718-972-6200   800-972-6201
JudaicaPr@aol.com
www.judaicapress.com

Printed in the United States of America

Lee M. Spetner received the PhD degree in physics from MIT in 1950. He was with the Applied Physics Laboratory of the Johns Hopkins University from 1951 to 1970, where he was engaged in research and development in signal processing and the scattering of electromagnetic waves from the earth's surface. From 1958 he was a member of the principal professional staff of the laboratory. He spent the academic year 1962-63 on a fellowship in the Department of Biophysics at the Johns Hopkins University. During that time he became interested in evolution and published several papers investigating information buildup in evolution. He taught graduate courses in physics at Howard University for about 5 years, and for about 10 years he taught information and communication theory at the Johns Hopkins University and at the Weizman Institute. In 1970 he moved to Israel where he took the position as technical director of Eljim Ltd., a new subsidiary of KMS Inc. of the US, engaged in research and development in military electronics. In 1976, Elbit Ltd., an Israeli company, bought Eljim, and Dr. Spetner continued with Elbit as operations manager of the Eljim facility. In 1984 he left Elbit to engage in private consulting until his retirement in 1992. From 1964, his hobby has been the study of organic evolution.

TABLE OF CONTENTS

PREFACE

CONVENTIONAL wisdom holds that life arose spontaneously. In the remote past a simple living organism is supposed to have formed by chance out of inert matter. That organism is then supposed to have reproduced and developed into the life of today through random variation shaped by natural selection. Although not yet widely recognized, the discoveries in biology during the past thirty or forty years, together with elementary principles of information theory, have made this view untenable.

A few biologists have pointed this out, but the majority seem to be unaware that there is a problem. The main problem, in my opinion, is the nearly universal belief among biologists that random variations shaped by natural selection produces large-scale evolution. According to this belief, random variation can lead to the large-scale evolution necessary for all the life of today to have developed from something even simpler than a single cell.

Information theory, which was introduced as a discipline about half a century ago by Claude Shannon, has thrown new light on this problem. It turns out that random variation cannot lead to large evolutionary changes. The information required for large-scale evolution cannot come from random variations. There are cases in which random mutations do lead to evolution on a small scale. But it turns out that, in these instances, no information is added to the organism. Most often, information is lost. A process that adds no heritable information into the organism cannot lead to the grand evolutionary advances envisioned by the neo-Darwinians. In this book I shall take you through an argument leading to this conclusion.

Why is randomness important? It is important because it has had a profound influence on the shaping of the *Weltanschauung* of Western Society. It has led to atheism and to the belief that we human beings are no more than a cosmic accident. This belief serves as a basis for the social values and morals held by Western intellectuals, and for their attitudes toward religion. If the belief is unfounded, then the resulting world view and its implications must be reexamined.

———

I have studied the implications of random variations on evolution for many years, and have found contradiction after contradiction, difficulty after difficulty, with neo-Darwinian theory. I gradually realized that massive evidence was pointing to an essential role for *nonrandomness* in evolution. It became clear to me that here may lie the answer to the puzzles and difficulties of evolution. This led me to propose an outline of a theory of evolution that accounts well for observable evolution, including the many instances where neo-Darwinian theory has difficulties.

A large body of evidence points to the suggestion that organisms come equipped with the ability to make heritable changes in their organs and functions in response to environmental cues. Much of this evidence has been available for well over a century. Furthermore, the idea I propose in this book has been proposed long ago, but it has been generally ignored. The increasing amount of data, accompanied by their increasing reliability and quantification, compels us now to give serious consideration to this proposal.

———

More than thirty years ago I suspected something was wrong with neo-Darwinian theory. I was then on a year's fellowship in the Johns Hopkins University Department of Biophysics, on leave from the Applied Physics Laboratory

(APL) of the University. I became interested in evolution during that year.

I received the PhD degree in physics from MIT in 1950, and joined APL in 1951. I spent most of my professional career doing research and development on information processing in electronic systems, and teaching information and communication theory. After I had been at APL for about a dozen years, I was offered a year's fellowship in the university's Department of Biophysics. There I was to solve problems in the extraction of signal from noise in DNA electron-micrographs. I accepted the fellowship and, as it turned out, I learned a lot about biology.

There, for the first time in my life, I met evolutionary theory in a serious way, and I found it hard to believe.[*] It clashed not only with my religious views, but also with my intuition about how the information in living organisms could have developed. I thought about these problems and published several papers on the subject in the professional literature between 1964 and 1970. I then put the subject aside and returned to my regular work. I, nevertheless, tried to keep up with the latest developments in molecular biology and genetics. I have done a great deal of reading in the past thirty years. From discussions I have had with biologists I have found that I have come to know more about evolution than do most biologists who have not specialized in it. The more I learned, the more I felt my views on evolution were vindicated and even strengthened.

In January 1987, in a hotel room in Singapore, I read Richard Dawkins's book *The Blind Watchmaker*. I could hardly contain myself while reading it because I had long ago

---

[*] The only previous formal study of biology that I had was in 1940 in high school in St. Louis. Now that I think of it, it seems to me that my biology teacher must not have believed in evolution because I don't think she ever mentioned it. The courses I took in biology at Johns Hopkins in the 1960's were a real eye-opener for me. Biology sure had changed between 1940 and 1963!

thought through his arguments and rejected them. I knew what was wrong with them. I felt I had to write a book setting forth my own views of evolution.

As I read, and thought, and wrote, I learned a lot. My views on evolution evolved. After I had developed my arguments, I discovered I was not alone. Some biologists had found the same difficulties with the neo-Darwinian theory that I had.

I decided the book for me to write should be for the layman. I thought the book might turn out to be controversial and might draw some fire from professional evolutionists. I, therefore, felt I must present my arguments fully (though clearly and simply) and, for backing, to give full references to the literature. This is that book. In citing the literature, I have preferred to cite articles accessible to the layman. I have preferred to cite articles in the more popular journals such as *Scientific American*. I have also, where possible, tried to cite review articles. As a last resort, I cited text books and articles in the general professional literature. I make no pretense that my citations serve to give credit to the scientists who made the original discoveries. This isn't that kind of book.

———

The *first chapter* is historical background, and the *second* gives some background in biology needed for the rest of the book. The *third chapter* describes the neo-Darwinian theory, and the *fourth* shows why random variation cannot produce the kind of evolution the neo-Darwinians intended to account for. The *fifth chapter* considers the question of building up information by evolution. In that chapter I present several examples of mutations discovered in the past several decades that evolutionists offer to demonstrate evolution. I show, for each example, that it is not the kind of mutation that could contribute to large-scale evolution. There are several known examples of mutations that lead to small evolutionary changes. But there are none that can serve as a prototype for a component step in macroevolution. Chapters 4 and 5 are the key chapters in my argument against the evolutionary effectiveness of random variations. The *sixth chapter* takes on Richard Dawkins's thesis in his book *The Blind Watchmaker*.

In the *seventh chapter*, I suggest how there could be evolution without randomness. The main idea is that the capacity to adapt to a variety of environments is built into the organism. The environment induces the expression of this capacity. Cues from the environment combine with the information in the genes to develop the form of the organism. In making this suggestion, I have drawn from ideas put forth by several biologists who hold that evolution cannot be built on randomness. I cite a wide sample of the large body of evidence for my suggestion. I cite several of the many examples where the environment affects the form of the organism and is heritable. The suggestion accounts for some of the very recent data that are disturbing to the neo-Darwinian theory. It also accounts for many well-known phenomena that have been said to be Lamarckian. But it accounts for them in a causal way, through known mechanisms.

The eighth chapter is the *Epilogue* in which I discuss the implications of the conclusion that evolution cannot be built on randomness. What effect would such a conclusion have on our world view? If evolution cannot be based on random variations, then it must come through a nonrandom mechanism. This mechanism might be something like what I have described in Chapter 7, or it may be some other nonrandom mechanism. The mechanism itself could not have evolved. Then how did the capacity for it ever arise? Science cannot now answer that question. Whether it will ever be able to answer it is a matter of conjecture. Is *Creation* an option?

————

Many people helped me in preparing this book. The conclusions, though, are mine and, of course, I am responsible for any errors. My thanks go to many, especially to the following who read the manuscript and gave me their comments. I thank Professors Yaacov Leshem and Sanford Sampson, both of the Life Sciences Department of Bar Ilan University, Professor Alvin Radkowsky of the Department of Nuclear Engineering of Tel Aviv University, the late Professor Christian Anfinsen ע״ה of the Department of Biology of The Johns Hopkins University, Professor Edward Simon of the Department of Biological Sciences of Purdue University, Dr. Harry Langbeheim of Sigma Chemical Corporation, Rehovot, Dr. Ron Wides of the Life Sciences Department of Bar Ilan University, Mr. Laurin Lewis of the Hebrew University, and Hilda Krumbein who edited some of the early drafts.

I also thank Pamela Aronson who read my later drafts and suggested many valuable improvements in style. I thank all of them for their encouragement and their criticism, even if some of it was sharp. Particular thanks go to two of my sons, Abba and Daniel, who have read carefully through many drafts. They checked all the mathematics (when I still had equations in the draft). They also made many good suggestions in mathematics, in microbiology, and in logic. I accepted most of the suggestions I got.

My wife Julie deserves special thanks for the sympathetic attitude she displayed throughout the ordeal of creating this book. Moreover, she did the printing and last-stage proofreading of the many drafts of the manuscript.

My thanks also go to those who found typographic and other errors that slipped by me in the first printing of the book, enabling me to correct them in this revised printing.

I hope you will read the book with an open and inquisitive mind, that you will follow my arguments and finally agree with my conclusion. I hope you enjoy reading it.

Rehovot
1997

Chapter 1

HISTORICAL BACKGROUND

IT was a cool cloudy day in London in late September of 1859 when John Murray took delivery of the author's final proofs of the manuscript. Murray's grandfather had started the publishing business, and John was determined to keep it successful. Although he was afraid of losing money by publishing what he thought was an unpromising book, the famous geologist, Sir Charles Lyell, pressed him into it. Lyell had intimidated him so, that John even raised his intended printing run from 500 to 1,250 copies — modest by today's standards, but risky for Murray's business. On publication day, to the surprise of everyone, he sold out all the copies at 15 shillings each. With his publication of Darwin's *Origin of Species*, John Murray modestly launched the greatest intellectual revolution of the modern world.

The *Origin of Species* radically changed the conventional wisdom of Western civilization. Until about a hundred and thirty years ago, conventional wisdom held man's origin to be supernatural. It held that all life was created by a Power Who made each form of life separately. To Western man that Power was the Creator and Ruler of the universe. With the rise of science and the scientific method, opinions slowly began to shift. Kepler, Copernicus, and Galileo showed that the earth was not the center of the universe. Newton showed that the earth, the moon, and the planets moved through the heavens under fixed laws; indeed, Newton's laws could explain the motion of bodies on earth as well as in the heavens.

As scientists began to see the universe run by fixed laws, man began to wonder if there is any need, or indeed any place, for a Divine Ruler. The fixed laws seemed to be enough to run the universe. One could wonder, is the universe run by a Divine Being, or does it run by itself? Are we humans the product of a Divine Creation set on this earth to fulfill some purpose? Or are we merely a product of the laws of nature, like a rock falling off a cliff? There were advantages to doing away with the need for a Divine Ruler: man would then be answerable only to himself and not to some Higher Authority.

But could one really do away with the need for a Creator? The universe had to have been created — or did it? Perhaps it was never created, but had existed for ever. That idea had Aristotelian backing.*

But even if the inanimate universe had existed forever, surely life had to have been created. Pasteur showed that life comes only from life. Life had to have a beginning. It couldn't have existed forever. Was life created? If it was, then man might have to live with the idea of a Divine Creator, Who would have the right and the ability to be his Ruler.

In his famous book, Darwin suggested that the great complexity of life could have developed in a natural way from some very simple living form. That allowed one to suppose that a sufficiently simple form of life might have formed from inanimate matter — again, in a natural way. Some of Darwin's followers went even further and suggested that there was no Creator, and therefore no Divine Ruler.

---

* Until the middle of the twentieth century, conventional wisdom held the eternity of the universe to be a "scientific fact." The eternity of the universe was eventually overthrown by Edwin Hubble's discovery in 1923 of the universal expansion, the "Big Bang" theory of the origin of the universe that stemmed from it, and finally by the discovery in 1965 by Arno Penzias and Robert Wilson of the background microwave radiation, considered to be a relic of the universe's creation.

Although the theory of evolution is often thought of as Darwin's creation, some of his ideas were around long before him. The notion that life has evolved has itself evolved from early ideas that date back at least to the ancient Greeks. Theories of Anaximander and Empedocles held that all animals, including man, began in water. According to these theories, some of the animals left the water and adapted to living on land. The Greeks even wrote of a kind of natural selection, which later formed the main point of Charles Darwin's theory. But these ideas were not blessed with Aristotle's backing, and they never gained a central place in ancient science. They remained outside the mainstream of science and philosophy.

The closest the philosophy of the Middle Ages got to evolution was in the concept of the *scale of nature*, which divided all of creation into four levels: *mineral, plant, animal,* and *man*. This division led some medieval thinkers to speculate about the origin of the higher forms of life. They speculated that the higher forms may have evolved from the lower ones, but unlike Darwin, they didn't explain how evolution might have come about. The natural selection of the Greeks also did not explain how evolution could work.

The eighteenth century saw the beginnings of the modern ideas of evolution. Yet the creationist position remained firmly in place among most scientists until the latter half of the nineteenth century. The creationist view of the science of living things carried the authority of Carolus Linnaeus (1707-1778). Linnaeus was a prominent Swedish physician and botanist. Born the son of a poor clergyman, he became the most influential biologist of his time. He held the opinion that species do not change and that all living things were created as they are today. On this basis he classified living things into those fixed groups that he identified with the descendants of the original forms made by the Creator.

3

Linnaeus set up the binomial, or two-fold, system of naming living things that biologists still use today as part of their classification system. Linnaeus's system classifies each organism into a group known as a *species*, and each species into a higher-order group called a *genus*.

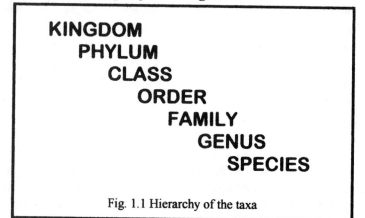

KINGDOM
PHYLUM
CLASS
ORDER
FAMILY
GENUS
SPECIES

Fig. 1.1 Hierarchy of the taxa

Our present system puts the genus under still higher-order groups, or *taxa*. Genera[*] that have much in common are put in the same family. Families make up an order; orders make up a class; classes make up a phylum. Phyla make up a kingdom. Fig. 1.1 shows the hierarchy of these taxa.

Linnaeus defined *species*, after the definition of John Ray (1627-1705), as a group whose members could interbreed. Although Linnaeus at first believed the species were the forms that were created, he later changed his mind and held that they were the genera instead.

Continuing in the footsteps of Linnaeus was the prominent French scientist Georges Cuvier (1769-1832), who wrote extensively on zoology, comparative anatomy, and paleontology. In his studies Cuvier discovered that the fossil vertebrates could be put into a sequence from fish to mammal. He did not think, however, that this sequence meant that one form descended from the other. He believed it was rather the

---

[*] Plural of *genus*.

4

result of a succession of creations, and that a series of catastrophes had wiped out many life forms. He is the epitome of the scientist of the pre-Darwin era whose religious views were strictly Biblical. Ironically, his scientific work played an important role in the establishment of the theory of evolution.

William Paley (1743-1805), an English Christian philosopher and author, wrote two influential books on the relation between religion and science. In these books he tried to show how one could find religious faith by observing nature. The first of the two, which he published in 1794, he called *A View of the Evidences of Christianity*. It was required reading for entrance to Cambridge University until well into the 20th century. In 1802 he published another book he called *Natural Theology*. This book continued the theme of the first, and in it he updated an old argument for the existence of a Creator, which has become known as the *Argument from Design*.

The Argument from Design says that a very complex object must have been designed, and therefore, there must have been a designer. Paley argued, "There cannot be design without a designer." He reasoned that if you were walking in the desert and found a watch on the ground you wouldn't suppose for even a moment that the watch had come into being by itself, by chance.* The balance and spring were first introduced as the controlling element of the watch toward the end of the seventeenth century. A watch is evidence of design; you would surely assume that someone must have designed and built it. How much more so must something as complex as a living organism have been designed and built by Someone. The argument concludes that there must have been a Creator of living things.†

---

*I would guess that Paley used a watch as his example because the watch was the "high technology" of the eighteenth century.

†Although Paley first introduced the English-speaking world to the *Argument from Design*, it did not originate with him. The argument is found in the *Midrash Temurah*, Chap. 3 (end), dating back two millennia and attributed to the Mishnaic sage Rabbi Akiva.

The strength of the Argument from Design kept evolution from becoming a serious issue until Darwin gave what seemed to be a good answer to it. But Darwin's answer was not good enough. It turns out that neither Darwin's theory of evolution, nor the more modern versions of it proposed to this day, are good enough to lay to rest the Argument from Design.

———

Before the middle of the nineteenth century, the majority of scientists accepted the Argument from Design; the Argument even had a strong effect on the young Darwin himself [Raven 1967]. As he was working out his theory, Darwin struggled with the question of how life could have arisen. He finally concluded that life could develop complexity without the need for a designer.

The earliest of the modern influential evolutionists was Georges Leclerc Buffon (1707-1788). Buffon was a well-known French naturalist. Born of well-to-do parents, he was trained in law, but was later attracted to science. He became an expert on timber, and was elected a Fellow of the Royal Society of London while on a trip there. Later, he was also elected to membership in the French Academy, and he served as its treasurer. The French Minister of Marines asked him to catalogue the king's museum. Buffon accepted the assignment and expanded the project into a general account of all of nature. His study resulted in a 44-volume *magnum opus* on natural history, which took more than 50 years to complete and publish. This work and other scientific achievements established his prestige as a scientist.

Buffon became one of the early advocates of evolution. He challenged Linnaeus both on his system of classification and on his view that species and genera don't change. Buffon held that living things evolved, that species would advance or regress as the environment changed. But he did not explain how these changes might occur, as Darwin was later to do.

A prominent follower of Buffon was the French naturalist Jean Baptiste Lamarck (1744-1829). Born into a family of soldiers, he was the youngest of eleven children. His father sent him to study for the priesthood; but when his father died, he left his studies to join the army. He became an officer in the French army and fought the Prussians in the Seven Years' War. He distinguished himself for bravery in the war, but he had to quit the army because of a neck injury. He then went to study medicine in Paris and became interested in meteorology and botany. He soon switched his studies from medicine to botany. Eventually, he became an expert in both botany and invertebrate animals and fossils.

Lamarck came under the influence of Buffon and later began to develop his own ideas on evolution. He proposed his own theory to explain how organisms might have developed from earlier ones. He laid down four laws to describe how animals could have developed and how a new organ might result from a new need of the animal that encourages a new type of behavior. The fourth of these laws says that traits an animal acquires can be inherited. His theory was ridiculed by the scientific establishment of his time, who believed in creation and the fixity of species. (His theory was later ridiculed by Darwin and his supporters as well.) A victim of failing eyesight for many years, Lamarck was blind at the end of his life and died in great poverty.

Even after Buffon and Lamarck, the evolutionary camp was weak; it was weak in numbers, in prestige, and in the influence it wielded. But by the first half of the 19th century, evolution began to gain momentum. A few scientists of repute began to think, talk, and write of the likelihood of evolution; and they lent the weight of their prestige to the evolution side of the controversy. One of these was Geoffry Saint-Hilaire (1772-1844), a French mineralogist and zoologist. He held that since all animals have related structures they must all be related. He also held that higher forms arose from lower ones. Other evolutionists of this period were the German geologist Christian von Buch (1774-1853), the Swiss botanist Pyrame de Candolle (1778-1841), and the German anatomist Lorenz Oken (1779-1851).

Charles Darwin (1809-1882) was unpromising as a student. He was described by his teachers as an "ordinary boy", and "rather below the common standard in intellect" [Himmelfarb 1962]. He found school a bore and looked on it as a waste of time. His father took him out of grammar school at the age of sixteen, a year before he was to have finished, and sent him to Edinburgh University to study medicine. His father, a doctor and the son of a doctor, hoped his son would follow him in the medical profession. But Charles's lack of interest in medicine made him leave Edinburgh after two years. He drifted for about a year, doing little of consequence; he seemed destined to have no proper career and to lead an idle life.

Not knowing what else to do with his lackluster son, his father sent him to Cambridge University at the age of nineteen to prepare for a career in the church. In his early years Charles was a religious boy, not knowing that his father was an unbeliever. Until he was in his late twenties, he looked forward to becoming a country clergyman. He finished Cambridge at the age of twenty-two and a half, without distinction, but with a deep interest in science. He was now a gentleman, a likable young man, six feet tall, healthy, good-natured, unambitious, and unassuming.

The turning point of his career came when he received an appointment as the naturalist on the scientific voyage around the world of the H.M.S. *Beagle*. The appointment carried no pay, but it afforded Darwin the opportunity to apply his talents to the study of nature.

He set sail less than two months before his twenty-third birthday. During the voyage he observed and classified plants and animals of many lands and collected many specimens. His accomplishments were all the more noteworthy because they were made in spite of his terrible attacks of seasickness, which did not abate throughout the entire voyage. On this voyage he laid the foundation of his life's work on the theory of evolution.

The voyage was to last two years, but ended up lasting five. Darwin returned home mature and filled with a wide knowledge of field geology, botany, zoology, and paleontology. He would later use much of this knowledge in building his theory. He spent the next few years writing the journal of his voyage, and the germ of his theory began to develop slowly in his mind.

Two years after returning from his voyage he married his first cousin, Emma Wedgwood. His father settled a sizable sum of money on him, and Emma also received a respectable dowry from her father. Darwin invested the money wisely and was able to live on the income. He could pursue his research without having to spend time earning a living. After a few years of living in London, he bought an eighteen-acre estate in the small village of Down, Kent, about twenty miles southeast of London, where he could work in the solitude he preferred.

His research covered a variety of topics, yet all the while his theory of evolution was percolating in his mind. He wrote several works before 1859 that stemmed from the *Beagle* voyage. These included volumes on geology, among which were one on coral reefs and one on volcanic islands. He also wrote on zoology. These works contained both his observations and his theorizing. His most extensive work during this period was a four-volume work on barnacles, which brought him the Royal-Society medal for biology in 1853.

Darwin's health was constantly bad. His health began to fail after returning from the voyage, and it was poor for the rest of his life.* Pain would keep him from putting in a full

---

* According to his grandson, Sir Charles Galton Darwin, the decline in his health was unrelated to his voyage on the *Beagle* [Darwin C. G. 1959].

day's work. If he worked four hours in a day, he considered himself to have done well. He would retire in the early evening, but pain usually kept him from sleeping through the night. His poor health did not permit him to attend scientific meetings or otherwise to travel far from his home in Down.

In 1856, at the urging of his friend, the geologist Sir Charles Lyell, he began writing a full account of his theory of evolution, which he planned to be an extensive work. By 1858, when it was about half done, he received a manuscript in the mail from an English naturalist, Alfred Russel Wallace, who was then exploring Malaysia. In his manuscript, Wallace described the very theory Darwin had in mind and was now writing up. Russel had sent it to Darwin for his opinion.

Darwin was now faced with a dilemma. How could he publish his work when Wallace had confided to him the same theory and had already prepared it for publication? Lyell suggested as a compromise that both men should publish their works together. Thus Wallace's paper, together with extracts of Darwin's sketch of his theory written in 1844, were presented jointly as papers to the Linnean Society.

The incident spurred Darwin to hurry to publish his full theory, which he did in his book of 1859, *On the Origin of Species by Means of Natural Selection,* or the *Preservation of Favoured Races in the Struggle for Life.* It is more popularly known as *The Origin of Species*; but for convenience, I shall refer to it simply as *The Origin.*

Darwin received many honors during his lifetime, not all of them in recognition of his scientific achievements. The city of Darwin, Australia, is named after him as is Mount Darwin near the southern tip of South America. Ironically, neither of these names had anything to do with his scientific work. The mountain got its name during the voyage of the *Beagle* when Captain Fitzroy named it in honor of Darwin's courageous act of saving their boats from destruction when the party was encamped under the mountain. The Australian port city got its name during the second voyage of the *Beagle* when Captain Wickham was charged with giving names to all the unnamed capes and bays he was exploring. To ease his job, he made a list of all his former shipmates on the *Beagle's* first voyage and used those names one by one. Darwin's name just happened to come up when he came to the bay on which Darwin, Australia, now sits [Darwin C. G. 1959].

———

To understand Darwin's work you have to distinguish between his *theory of descent* and his *theory of natural selection*. The full name of the first is the *theory of descent with modification*. Some call it the *fact of evolution*, and some call it the *doctrine of evolution*. Darwin's theory of natural selection was his explanation of how descent works.

Darwin intended his theory of descent to be a conclusion drawn from his theory of natural selection. The theory of descent says that all living organisms arose from a single, simple form. They all developed in a natural way into their present forms. Darwin held natural selection to be the mechanism through which evolution works.

As I noted earlier, many people had suggested ideas about evolution long before Darwin wrote his book, but none was widely accepted [Himmelfarb 1962]. In contrast, within ten years of its publication, Darwin's work met with broad acceptance. Scientists accepted as fact the idea that life evolved. Most Western thinkers followed; and they, too, accepted the main point of Darwin's theory.

Just what was in Darwin's theory that enabled it to overcome the force of the Argument from Design? What argument did he have that could convert the strong opposition to evolution into strong support? What was there in his thesis that had enough force to change a climate against evolution into one of wide acceptance? The Argument from Design, after all, seemed so plausible. Surely, you wouldn't think a watch could have come together by chance; the chance of such a thing happening is so small as to seem wholly unreasonable. Then shouldn't the same argument hold even more so for a pigeon or a turtle?

Darwin's answer was that Paley's argument for the watch cannot be used for the pigeon. Although the suggestion about the watch is indeed unreasonable, that about the pigeon may well be sound. The pigeon could have come about by natural means. A pigeon differs from a watch in a way that's important for Darwin's argument: the pigeon reproduces itself, but the watch does not.* Darwin didn't suggest that a pigeon ever came into being all at once, but he did suggest that it developed gradually from a much simpler form of life. He thought of that simple form as a single cell. Pigeons, or plants and animals generally, could arise gradually through a long series of small steps. A watch, however, would have to come into being all at once. The chance of either the pigeon or the watch coming together all at once is small enough to be negligible. Dividing the watch's assembly into stages won't help its chance of success, but dividing the pigeon's assembly into stages might help a lot.

Darwin saw pigeon breeders develop many new and strange features in their pigeons. Breeders would exploit the natural variation in a brood of pigeons, and choose what pleased their fancy. In each brood they would select for further breeding the pigeons that showed a change in the direction they wanted.

Darwin suggested that, given enough time, there was no limit to what one could achieve by selective breeding. He hit on the idea that the ordinary workings of nature could play the role of the pigeon breeder. Nature would do the selecting! An animal born with a heritable advantage, no matter how slight, would tend to have more offspring. The offspring would also carry the advantage, and they would, in turn, re-

---

* Paley did discuss what the finder might think if he discovered that the watch could replicate itself. But he did not think of Darwin's argument of natural variation and natural selection which, according to Darwin's followers, would tend to invalidate Paley's conclusion.

produce. Darwin thought the effect of natural selection could be even stronger than that of artificial selection by humans, if it acted over a long enough time.

Darwin's theory of natural selection is the following. There are many small variations already in the population, and new ones are appearing all the time. He assumed these variations are heritable. An organism whose heritable traits give it an advantage will have offspring with the same traits and the same advantage. If the advantage helps it survive better and reproduce more than the others, then its offspring will tend to increase more than others.

In wild populations, natural forces play a role much like the role the human breeder plays with the pigeons: the breeder selects those pigeons he likes for further breeding. So too, nature "selects" those animals with an advantageous trait. Those that have a trait giving them more surviving offspring will pass that trait on to their progeny, who will in turn also have more surviving offspring, and so on. In this way the advantage builds up in the population.

The animals with such an advantage would increase in number, much as a bank account grows with compound interest. Those without the advantage would not build up so rapidly; their number would even drop if the population were limited. In fact, populations usually are limited. They are limited by such factors as space and food supply. As those with the advantage would increase, they would become more than just a majority in the population — they would eventually become the whole of it. Darwin called this process *natural selection*.

———

In the formulation of his theory, Darwin was heavily influenced by the writings of Thomas Robert Malthus (1766-1834). Malthus was a clergyman and a professor of history and political economy. In his book on population he noted that a population tends to increase exponentially while the means of production increase only linearly. The population size, therefore, is always destined to outstrip the means of production. He saw that the human population could be limited only by vice (by which Malthus meant birth control), famine, war, and disease. The human population is, therefore, inevitably driven to hover constantly on the edge of misery and starvation.

Malthus's book made a profound impression on Darwin. Darwin wrote:

> In October (1838), that is fifteen months after I had begun my systematic inquiry, I happened to read for amusement "Malthus on Population," and being well prepared to appreciate the struggle for existence which everywhere goes on from long continued observation of the habits of animals and plants, it at once struck me that under these circumstances favorable variations would tend to be preserved, and unfavorable ones to be destroyed. The result of this would be the formation of new species. Here then I had at last got a theory by which to work. [Eliot 1909, p. 7]

Darwin drew an analogy from Malthus's view of the human population to animal populations in the wild. He saw animal populations constrained by the same kind of limitations Malthus ascribed to human populations; he saw those limitations creating a "struggle for existence" which formed the rationale for his mechanism of evolution.

Under ideal conditions living things tend to increase their numbers exponentially. A single bacterium, if given unlimited nourishment and space would, after 24 hours produce about a kilogram of bacteria. If the ideal conditions could be kept up for another 24 hours, there would be 300 billion tons of them! Plants and animals also tend to increase exponentially. The same numbers could apply to any plant or animal as with bacteria, but it would take a little longer. Of course, long before populations could reach unreasonably large numbers, food and living space would run out, and conditions would cease to be "ideal."

Predator animals do not overexploit prey populations. They live only on the income from their resources and preserve the principal for posterity.

Darwin suggested that living things strongly compete when their populations rise to exceed their resources. This competition, according to him, leads to a natural selection of the most fit. Under the miserable conditions of not enough food or living space, only the hardiest individuals would survive and procreate.

Later discoveries, however, have shown that animal populations do not follow Malthus's prognosis for the human race. Darwin erred in the insight that led him to his theory of evolution. Animal populations generally do not hug the brink of disaster. Population size is not controlled by starvation, disease, or predation [Wynne-Edwards 1986]. In most animal populations mass starvation and disease are rare; they occur only as the result of extraordinary catastrophes, such as droughts or epidemics. Moreover, predator animals do not overexploit prey populations. Man seems to be the only predator that overexploits its resources and drives them to extinction. Animal predators generally take only the excess of the prey population [Wynne-Edwards 1986]. They live, so to speak, only on the income from their resources and preserve the principal for posterity.

Wynne-Edwards [1986] has suggested that animals generally manage their food resources by controlling their own numbers. Populations are kept in check not by the extrinsic forces of mass starvation or disease, but by intrinsic forces built into the animals themselves. This phenomenon may be surprising and even amazing to most people, but biologists studying animals in the wild have reported this kind of control operating in a variety of populations [Wynne-Edwards 1965, 1986; Bergerud 1983].

Plants also do not proliferate in a field to the point where they become crowded. They do not engage in a "struggle for existence" where natural selection would preserve the strong and destroy the weak. Plants tend to control their populations by sensing the density of the planting. When the growth is dense, plants produce less seeds; when growth is thin, they produce more seeds [Bradshaw 1965].

————

Darwin's process is supposed to take a long series of steps. He suggested there would be many steps with plenty of time for each. Each would build on the one before it. They would accumulate and lead to a large change in the population. In principle, Darwin expected the process to lead ultimately from a single, simple cell to complex forms like pigeons and people. Large changes, he said, could be attributed to the *cumulative* action *of natural selection*. He compared it to the deliberate breeding of plants and animals where the key to the large changes lies in "man's power of *accumulative* selection."*

He admittedly did not know the source of the variations he required, but he was sure they were there. He was hesitant about labeling them as random, even though he had referred to them as having occurred by "chance." He explained that his use of the word "chance" was only to indicate that we do not know their causes [Darwin 1873, pp. 128 ff.].

He did not refrain from speculating on their causes, however. He felt there *were* definite causes for the variations. They could be brought on by environmental conditions or they could be brought on by the use or disuse of organs. But he did not want to call them random.

————

Darwin's attack on the Argument from Design seemed successful. A pigeon reproduces itself, but a watch does not. The difference seemed enough to blunt the force of the Argument from Design for living and reproducing organisms. He sidestepped the Argument from Design by suggesting that evolution works by the method of *many small changes.*

——————————————

* Richard Dawkins [1986] used Darwin's term *cumulative selection* as a descriptive term for this chain of evolutionary steps, with selection acting at each step.

That's how he thought he could explain how life may have developed in a natural way. With the Argument from Design out of the way, scientists could accept evolution.

Darwin based the theory of descent on his theory of natural selection, the key point of *The Origin*. *The Origin* described what seemed to many to be a reasonable way for descent to result from natural laws. The scientific world accepted descent only after Darwin gave them his theory of natural selection. As Darwin saw it, the purpose of natural selection was to explain how descent might work.

———

Darwin led the world into thinking about the evolution of living things; almost all of biology has followed his lead. He opened a whole new field of scientific study that would, in time, touch upon every branch of the life sciences.

There were, however, some serious problems with evolution. Darwin and his contemporaries assumed that the traits of the parents blended in their offspring. A tall father and a short mother, they assumed, would lead to a daughter of medium height. Critics of the Darwinian theory said that any good variation that might arise would gradually disappear; they thought it would be diluted more and more as generations went on; they thought it would not remain in the population long enough for natural selection to get a hold on it.

That problem was solved just six years after Darwin's book was published, but neither Darwin nor his colleagues knew about it at the time. It was solved by a monk named Gregor Mendel working alone and doing his research quietly in a monastery.

Gregor Mendel (1822-1884) was the son of a poor Austrian farmer. He spent his early years working on his father's farm, where he developed an interest in plants. At the age of 21 he entered the Augustinian monastery in Brünn and was ordained a priest four years later. His abbot sent him to study science and mathematics at the University of Vienna for a few years. After his return, he taught science in nearby schools.

He taught for about fourteen years, until he was elected abbot of the monastery in 1868. While he was teaching he began, in 1856, to experiment with plants in the small monastery garden. He spent his time in the garden studying what happened when he crossbred peas. After he became abbot, he no longer had the time to pursue his science, but he retained a strong interest in it for the rest of his life. In 1865 he published his important discovery that would in time give the support that Darwin's theory so badly needed.

Mendel found that inherited traits do not get mixed and diluted. He found that the offspring of a yellow pea and a green pea is not some mixture of yellow and green; it's either yellow or green. A hybrid cross between a pure-bred yellow and a pure-bred green is yellow. Yet the yellow hybrid plant can produce a green pea if it were fertilized by a green-pea plant, or by even another yellow hybrid.

Mendel showed that the traits inherited from parents do not blend and average out — they retain their full potential purity. A blue-eyed mother and a brown-eyed father, for example, do not have a son with bluish brown eyes; he will have either blue or brown eyes.*

Darwin never knew of Mendel or his work; Mendel published his work in an obscure journal. The scientific community discovered his work only in 1900, eighteen years after Darwin died, when a Dutch botanist, Hugo de Vries (1848-1935), brought it to their attention. De Vries took an interest in Mendel's results, and he checked them by doing his own experiments. He also found that new, heritable variations can appear suddenly, and he called these variations *mutations*. But de Vries and others found that mutations were not the

---

* If the parents are each pure bred, the offspring will all be hybrids and have brown eyes. If the father is a hybrid, then there is an even chance an offspring will have blue, rather than brown, eyes.

kind of variations Darwin described or needed. Mutations are usually harmful to the organism. They are never useful, and are often lethal. So scientists didn't think mutations were suitable for Darwin's theory; they thought these mutations could not be the source of the variation needed by the Darwinian theory.

At the start of the 20th century, there were other suggestions for theories of evolution. Some wanted to get rid of natural selection and explain descent in other ways. But none of those theories became widely held, and none of them ever replaced the Darwinian theory.

By the first third of the twentieth century, Darwin's theory had been overtaken by new discoveries in biology. These discoveries were mainly in the new science of genetics, which include both *transmission genetics* and *population genetics*. The former is concerned with individuals and how their genes pass on to their descendants. The latter is concerned with populations and how their gene frequencies change. By the end of the 1930's the theory was in disarray. Unanswered riddles haunted the theory, and it was in serious need of repair and updating.

The subject was raised at a meeting of the Geological Society of America in 1941. A suggestion was made there that geneticists join hands with morphologists, taxonomists, and paleontologists to try and come up with a modernized version of Darwin's theory. They were asked to try to rebuild the theory using the latest findings in all these fields. Specialists responded to the call, and over the next few years produced a synthesis of their fields into a revised theory of evolution.

The scientists who participated in establishing the new theory included the geneticists, G. Ledyard Stebbins and Theodosius Dobzhansky, the zoologists Ernst Mayr and Julian Huxley, the paleontologists George Gaylord Simpson and Glen L. Jepsen, and the mathematical geneticists Sir Ronald A. Fisher and Sewall Wright. They called this theory *the Modern Synthetic Theory of Evolution*. The theory gradually became known as the *Neo-Darwinian Theory of Evolution*, and its framers and adherents became known as *neo-Darwinians*. Their agenda called for a theory that could explain the development of life in a natural way. If they could account for the development of all the present complexity of life from some sufficiently simple first organism, it would help pave the way for a theory of a fully natural account of the actual origin of life.

The neo-Darwinian theory rejected Darwin's suggestion of the environmental induction of heritable variations and even more emphatically rejected the inheritance of acquired characteristics. Genes had by now been discovered, and although their molecular structure was still unknown, the neo-Darwinians had accepted the separation of the somatic and the germ cells as suggested half a century earlier by Weismann [1893]. It seemed clear to them that neither environmental influences nor acquired characteristics could affect the germ cells, and heritable variation could only stem from changes in the germ cells.

The neo-Darwinians were unwilling to accept environmental influence as a cause of variation, and they were unable to find a mechanism that could directly produce the genetic changes needed for descent. They therefore chose randomness as the source of the variations they needed — the very randomness that Darwin had rejected. Some of the variations are detrimental to the organism, but there may be some that are beneficial. A heritable variation of the latter kind, even if rare, will increase its number in the population by natural selection. Eventually, the change, if it enhances the organism's ability to survive and reproduce, will spread throughout the population.

The neo-Darwinians then built their theory on random variation culled and directed by natural selection. They identified the heritable variations needed by the theory with the *mutations* discovered and named by De Vries some forty years earlier. A decade later, Watson and Crick identified the

21

heritable variation with the random errors in DNA replication.

If the neo-Darwinian agenda had worked out, there would be no place for a Creator in the origin of life except to establish the laws by which the evolution had taken place. Even that position would not be an honorable one if the appearance of man were not inevitable, as Stephen J. Gould of Harvard University believes it is not [Gould 1989, pp. 292 ff.].

Chapter 2

INFORMATION AND LIFE

HOW does an acorn know it has to grow into an oak tree and not into a sunflower? How does a chicken egg know it has to grow into a chicken and not into a duck? More than that, how does the egg know *how* to grow into a chicken? Where does it get the information?

The science of biology took a pivotal turn about 40 years ago when biologists began to learn how information plays its role in living organisms. We have discovered the location of the information in the organism that tells it how to function and how to grow, how to live and how to reproduce. The information is in the seed as well as in the plant; it's in the egg as well as in the chicken. The egg passes the information to the chicken it becomes, and the chicken passes it to the egg it lays, and so on.

Evolutionists claim to know how it all began and how chickens and eggs became what they are today. They tell us that the neo-Darwinian theory describes how life evolved. But it turns out that the theory cannot account for the way information would have had to build up to make evolution work.

If we are to understand how life evolved from some simple form, we have to understand how the information of life could have evolved. Scientists have discovered where life's information is stored; they have learned a great deal about the message it contains and how that message is used. This chapter will give you some of the general background about information in living things. You will find more detailed background in notes at the end of the chapter and in the Appendix.

Life is so diversified that to any statement I could make about living organisms there are exceptions. Because of the many exceptions, I should qualify everything I say with hedging phrases such as "generally," "usually," and "almost always." But I'm afraid the constant repetition of these hedges will slow me down and bore you. So let's make a pact now that I forego the hedging phrases and you are to understand that almost all* my statements may have rare exceptions.

———

Single cells are the smallest and simplest of living things that can reproduce themselves. Bacteria and yeast are examples of living cells. A bacterium is a single cell that measures about one micrometer† across, and a yeast cell is about 5 micrometers across. The bacterial cells are so small that a trillion of them could fit into a teaspoon. Yet it takes a lot of information to define a bacterium, and all of it is tucked into only a small fraction of the cell's tiny volume.

Cells reproduce by dividing, making two or more cells from one.‡ The cell before division is called the *mother* cell, and the two cells it becomes after division are called *daughter* cells. Each daughter cell has within it all the information the mother cell had and all the tools it needs to function. Before a bacterium divides, it almost doubles its size. It duplicates its information, and it gives a copy to each daughter

---

* Oops!

† A micrometer is a millionth of a meter, or a thousandth of a millimeter. About twenty-five thousand of them make an inch.

‡ Usually a cell reproduces by making only two cells from one. But some protozoa, like malaria, reproduce through a process called *schizogony*. In this process, many, sometimes hundreds, of nuclei form first without the cell dividing. The nuclei then escape from the cell, each with some cytoplasm, and become cells of their own.

cell. Under the best conditions bacteria can divide every half hour.

———

A living organism made of more than one cell is called *multicellular*. It is an organized group of cells that work together as a unit. All plants and animals are multicellular.* A typical cell of such a unit measures between 10 and 30 micrometers across. They are larger than bacteria or yeast, and they also hold more information.

The cells of a multicellular organism are specialized; each has its own job. There are, for example, skin cells, nerve cells, and liver cells. Each type of cell has its job and helps in its own way to promote the well-being of the whole organism. The cells cooperate with each other, enabling the organism as a whole to live and function efficiently.

The reproduction of a multicellular organism is more complicated than that of a single cell. A multicellular organism starts as a single cell and builds an organization of many cells. From the single starting cell, the cells divide many times. At each division, one cell becomes two. Through these divisions the number of cells in a single organism can grow to be quite large. The human body, for example, has about 50 trillion cells. (This number corresponds to about 46 doublings.)

Most living things we know, both plants and animals, are multicellular. They can reproduce either sexually from two parents or asexually from a single parent. Plants often alternate between these two ways of reproducing. Some simple animals also can reproduce in both ways, but the more complex ones do so only sexually.

———

* This is according to the modern definition of plants and animals. The protozoa, which used to be called "single-celled animals" are now classed in the Kingdom Protoctista. Yeast, which used to be classed with plants, are now placed in the phylum Ascomycota of the Kingdom Fungi [Margulis and Schwartz 1988].

When an animal reproduces sexually, two parents make a child. Each parent contributes one cell, known as a sex cell, or *gamete*, and these two cells fuse into one. The mother contributes a female gamete (an egg), the father a male gamete (a sperm). The process through which an organism builds itself from a single cell is called *development*.

Development starts when the two gametes unite and fuse into a single cell, called the *zygote*. This fusion marks the beginning of a new organism. From that stage until birth the organism is an *embryo*.

The cells of the embryo divide, and their number doubles at each division. At an early stage the cells begin to become different from each other; we say they *differentiate*. In the frog this happens after the cells number a few thousand. Some animal embryos start to differentiate after just a few divisions [Griffiths et al. 1993]. As the embryo grows, the cells differentiate into all the tissues and organs of the body.

The cells change and develop according to a program that's part of the information built into each of them. The program also accepts inputs from outside the cell. We can compare the development program to one that runs a computer. All the various kinds of cells, such as skin cells, muscle cells, nerve cells, and blood cells, form by differentiation. We see that, according to evolutionary theory, if a fish has evolved into a salamander, then the program that builds a fish must have evolved into one that builds a salamander.

The development program doesn't stop at birth but continues to control the animal's growth and maintenance even after birth. The program continues to operate as long as the animal is alive. It leads not only to birth but beyond. It brings a young mammal to maturity, and maintains its body. The program dictates the building of new tissues as old ones are damaged or wear out. Some animals can even regenerate destroyed limbs. Many biologists consider this maintenance process as an extension of the development program.

———

The information in a cell plays a role much like that played by information in a factory. The production file in a factory contains a set of instructions that tell what each worker has to do at each stage. The production file is information carried by printed symbols; the developmental instructions in the cell are information carried by molecular symbols.

Most of the information in the cell is found on small bodies called chromosomes.* When the cell is ready to divide, the chromosomes are visible through a microscope, and they look like thin threads. In cells that have nuclei (which includes all organisms except bacteria†), the chromosomes are in the cell nucleus. The part of the chromosome that carries the information is a molecule known as *deoxyribonucleic acid*, or *DNA* for short. All the DNA in all the chromosomes of a cell is called the *genome*, and the information it contains is called *genetic information*.

In organisms that reproduce sexually, the chromosomes come in pairs; the members of each pair come one from each parent. Different species generally have different numbers of chromosomes; humans have a set of 23 pairs; the vinegar fly, Drosophila *melanogaster*, has four. Both members of a pair of chromosomes have the same basic information, although they are usually not identical.

Don't get the idea that a higher organism necessarily has more chromosomes than a lower one.‡ Chromosomes are

---

* A small part of the heritable information is in other parts of the cell, but by far, most of it is in the chromosomes.

† Again in keeping with modern usage, I am including what have been called blue-green algae with the bacteria. Blue-green algae are now called *cyanobacteria* and, like other bacteria, have no nuclei. True algae do have nuclei [Margulis and Schwartz 1988].

‡ I make no apologies for calling a human being a "higher" animal than the silkworm. Conventional political correctness tends to make some biologists go out of their way to avoid scaling living things into "higher" and "lower" forms. They try to avoid any value judgment by putting ourselves at the top. This avoidance often leads to awkward and unnecessary circumlocutions; I prefer to avoid the tyranny of po-

only a way of packaging the DNA, and their number is not a good indication of organism complexity. Man, for example, has 23 pairs of chromosomes, but the silk worm has 28, the dog has 39, and the carp has 52. The African trypanosome, a single-celled parasite that carries sleeping sickness, has hundreds of chromosomes [Donelson and Turner 1985].

The chromosomes of some organisms may have much more DNA than are in the chromosomes of others. You might then think the amount of DNA in the genome is a better way to measure organ complexity, but that's not entirely correct either. Although humans have 30 times the DNA of some insects, there are insects that have more than double the DNA in humans. The amount of DNA is not a reliable measure of complexity because not all the DNA may have to do with complexity; part of a genome may be just many repeats of the same section.

One of the pairs in a set of chromosomes determines the organism's sex. The members of this pair can be of two kinds. In *Drosophila*, as well as in many other species, one member of the pair is straight and is called the X chromosome; the other has a bent shape and is called Y. In *Drosophila* and in mammals the pair of sex chromosomes in the female is two X's; the male pair is an X and a Y. In birds and in some insects it's the other way around: the male has two of the same kind of sex chromosomes, and the female has two different ones. In these cases, the chromosomes are called Z and W instead of X and Y.

The DNA is a *polymer*, which means it's a large molecule built from many similar small molecular units. A polymer is like a chain, and the small units are its links. In the DNA the links are called *nucleotides*, which are distinguished from each other by their *bases*.[1*] There are four kinds of bases in

(..continued)
litical correctness. When silk worms start classifying living things, they are, for my part, welcome to put themselves at the top of the ladder.
* Superscript numbers refer to notes at the end of each chapter.

DNA. They're called *adenine, guanine, thymine,* and *cytosine.* As is customary, I shall refer to these bases by their initials A, G, T, and C. There are then four kinds of nucleotides, each having one of the four bases.*

James Watson and Francis Crick received the Nobel Prize in 1962 for discovering the structure of the DNA molecule. They showed it was like a two-stranded ribbon wound as a double helix [Watson and Crick 1953a, 1953b].

There is no chemical restriction on the order of the bases along a strand of DNA; the order can be anything at all. The order of the bases is then free to carry information.

The string of bases, along either of the two strands, is like a string of symbols in a message. As a string of letters of an alphabet can carry a message, so can a string of bases. The bases of the two strands are paired in a complementary way (see Appendix A). The sequence of bases in one strand then determines the sequence in the other. The same message is in both strands.† You can think of the *base pairs* (or bp's) as the symbols in the alphabet, or you can think of the nucleotides as the symbols. You might think of the DNA molecule as a tape on which the cell's information is recorded.

The sequence of bp's of the DNA is a message written in an alphabet of four symbols, and it holds the information of the cell. The genome contains the message that, among other things, tells how to build the organism. The message is copied at each cell division.

How did that message get written in the first place? The standard answer of the biologist today is that the message got written by itself, through evolution, and that evolution works the way the neo-Darwinian theory says it does. But I shall show that evolution cannot work that way.

---

* See Appendix A for more information on the structure of DNA.
† Don't confuse the two strands of the DNA molecule with the pair of chromosomes. Each chromosome has its own double-stranded DNA molecule.

Then how could the message have been written? Does it mean that life was indeed created the way conventional wisdom held before Darwin? Does it mean the message had to have been written by a Creator? Science cannot answer that question.

The genome can hold a lot of information. The genome of a bacterium, for example, is a string of a few million symbols. The genome of a mammal has from two to four billion. If you were to print those symbols in a book in ordinary type, the book for a bacterium would have about a thousand pages. The symbols for a mammal would fill two thousand volumes — enough to take up a library shelf the length of a football field! All this information is in the tiny chromosomes of each cell.*

The DNA in a single human chromosome, if it were straightened, could be as much as 10 centimeters long. Not all the chromosomes have this much DNA, but all together the DNA in the nucleus of a single human cell would be more than a meter long. If all the copies of the DNA in all the cells of your body were straightened and laid end to end they would be about 50 billion kilometers long! That's long enough to reach from the earth to beyond our solar system! Light would take about two days to travel the length of all the DNA molecules in your body!†

The DNA molecule has been called "the fundamental molecule of life" [Weinberg 1985]. In the last 30 or 40 years, molecular biologists have discovered a lot about how the cell uses its information [Stryer 1988].

———

If the DNA is like a tape with information on it, how does the cell play the tape? How does the cell translate the DNA information into action? What *is* the action? The action con-

———

* More on information storage in Appendix B.
† Of course, the length of all the DNA in your body is not a measure of information. The cells of your body all have the same information. I just think that the enormous length of the DNA is of some interest.

sists of using this information directly to make little devices. These devices are machines and structural elements in the form of single molecules. These molecular devices are called *protein* and *RNA*. The cell also makes more DNA when it copies the chromosomes.

Protein molecules are also polymers. They are chains whose links are small molecules of from 10 to 30 atoms each, known as *amino acids*. There are twenty kinds of amino acids that make up all the proteins in living things. The amino acids are joined as links in a chain to form the protein molecule. The amino acids have side chains that jut out from the backbone of the main protein chain. You can think of a protein as something like a girl's charm bracelet. The side chains play the role of the charms. There are twenty kinds of them, but a protein can have many of each kind. A typical protein will have a few hundred amino acids. The character of the side chains, and the order in which they occur, determine the properties of a protein.

The order of the amino acids in a protein determines its function and whether indeed it will have a function at all. If a protein is made from a random order of amino acids, the chances are nearly certain that it won't do anything. To make a protein that will do something useful, the cell has to get the right amino acids in the right order. The order of the amino acids has to be just right to give the protein the right three-dimensional shape and the right electric-charge distribution to make it do a job. With the amino acids in the proper order, the protein will have special and complex properties.

The cell performs thousands of different chemical reactions. Each reaction consists of changing a molecule into one or more others. All the chemical reactions in a cell are mediated by *catalysts*. A catalyst always comes out of a reaction unchanged, and it can be reused indefinitely. The catalyst acts on the molecule that is the input to the reaction, and produces the output molecule(s). The input is known as the *substrate*, and the output is known as the *product*.

The protein's most widespread role is as a catalyst in bio-chemical reactions, and in this role it is called an *enzyme*. An enzyme often speeds up chemical activity so much that it can make a reaction go that otherwise wouldn't. Each reaction has its own enzyme. An enzyme speeds up a reaction rate by a factor of at least a million [Darnell et al. 1986]. An increase in rate by factors of ten billion to a hundred trillion are not uncommon [Kraut 1988].[*] A factor of a hundred trillion means that what takes a thousandth of a second with the en-zyme would take about 3000 years without it. Most bio-chemical reactions would take so long without their enzyme that, in effect, they wouldn't go at all. Because enzymes con-trol nearly all chemical reactions in the cell, we can say that, to a large extent, proteins control the chemistry of life [Stryer 1988].[†]

――――――

The action of an enzyme is under the control of the cell. The cell can turn its enzymes ON and OFF as it needs them or not. If there is a need for the product, the enzyme is turned ON; when there is no longer a need for the product, the en-zyme is shut OFF. This control helps promote efficiency within the cell. Efficiency would not be served if the cell were to use up energy and substrate to make a product it doesn't need.

In one kind of control, the presence of the product of a re-action inhibits the enzyme that catalyzes the reaction. Control of this kind is a form of negative feedback, and is often called *feedback inhibition*. This kind of control enables the cell to husband its resources.[‡]

――――――――――――――――

[*] Rate enhancements of biochemical reactions by some enzymes may be even larger than this, but it's hard to say because such rates are hard to measure [Kraut 1988].
[†] For a brief account of how enzymes work, see Appendix C.
[‡] See Appendix D for some detail of how the cell controls enzyme ac-tivity.

Aside from their role as enzymes, there are proteins that fulfill other functions as well:

- There are proteins that determine the shape and structure of the organism. They make up the structural material that gives strength and support to cells and tissues. Proteins make up the cytoskeleton, or cell skeleton, that serves as the framework giving shape to the cell. Also, a protein known as *collagen* is the main component of the structural members of larger organisms; it gives mechanical strength to bone and skin [Darnell et al. 1986].

- They help control some of the vital cell functions. Some proteins help regulate the making of other proteins, and some play a role in controlling the development of the embryo as dictated by the genetic information.

- Proteins also play a role in the selective transport of small molecules. For example, hemoglobin is a protein that transports oxygen through the blood, from the lungs to each cell in the body.* Other proteins move ions and sugars across the cell membrane and into the cell. They bring in nutrients and expel wastes and toxic materials [Darnell et al. 1986].

- Proteins are the motor molecules, and are the source of motion in living organism. In muscle, two kinds of protein filaments are made to move across each other to produce muscle contraction. Muscle proteins convert chemical energy into mechanical energy. In single-celled organisms, the motion generated by proteins is used to move flagella. Proteins also provide the

---

* The hemoglobin in the blood carries some 50 times the amount of oxygen that could be dissolved in the blood without the aid of hemoglobin.

force to move the chromosomes when the cell divides [Stryer 1988].

- Proteins perform a major role in the transfer of information inside living organisms. Some proteins carry messages from cell to cell and to their proper destinations within the cell [Linder and Gilman 1992].

- Some proteins help to generate, transmit, and receive electrical nerve impulses in animals [Darnell et al. 1986].

- Some play a crucial role in the body's immune system. Antibodies are proteins that recognize foreign substances, combine with them, and mark them for destruction [Darnell et al. 1986].

The information in the genome tells the cell what kind of proteins to make. Because proteins play a dominant role in cell function, they play a dominant role in the whole organism. The information in the genome, by controlling the making of protein, fixes the form and function of the entire organism.

RNA or ribonucleic acid is a molecule resembling DNA in structure and, like DNA, it also carries information. The RNA molecule is also a polymer like DNA, but it's a single strand instead of a double one. RNA differs from DNA in two other respects as well. First, the backbone of the RNA molecule contains *ribose* sugar in contrast with the *deoxyribose* sugar of DNA. Second, in place of the thymine (T) base in DNA, the RNA has uracil (U). The four bases (or alphabet symbols) of RNA are then A, G, C, and U. Contrast this with the A, G, C, and T of DNA. As the A can pair with a T, so it can pair with a U.

The deoxyribose sugar of the DNA is much more stable than the ribose sugar of the RNA. That's why the cell uses DNA to store genetic information over a long term. The cell uses DNA to hold information for the life of the organism, and on for generations. It uses RNA, on the other hand, to carry information for shorter terms.

_____

A *gene* is a genetic unit, which geneticists have shown to be part of a chromosome. You can think of each chromosome as a string of genes. Geneticists once defined a gene as a part of the chromosome that influences some trait of the organism such as the shape of the wing of a vinegar fly or its eye color. They once thought genes to be strung like beads along the chromosome. Molecular geneticists have now probed the fine structure of the gene and have found it to be part of the sequence of nucleotides of the DNA molecule.

As we learn more and more about the function of the gene on the molecular level, its definition tends to get more and more complex. A gene can be defined accurately and simply, but not very informatively, as a functional region of the DNA molecule [Griffiths et al. 1993]. A more informative definition of a gene is all the DNA sequences necessary to produce a protein or an RNA molecule [Darnell et al. 1986]. Genes fall into two broad classes:

1. *Structural* genes encode a protein, which in turn may play either a structural or a catalytic role in the cell. Some genes encode RNA, which can also play either a structural or catalytic role.

2. *Regulatory* genes control what the cell does and even how it does it. One of their main jobs is to control which proteins, and how much of them, the cell makes. They often do this by encoding a *protein* that turns specific genes ON or OFF [Darnell et al. 1986].

The information in the DNA lies in large measure in the plans it contains for making protein. The information in the DNA dictates the order of the amino acids in each protein. The information in a gene is read and transcribed onto a special RNA molecule called *messenger RNA* (mRNA).[*] The transcription is mediated by a special enzyme called *RNA polymerase*.[†] The cell sends the mRNA to where protein is made. Protein is made by bodies in the cell called *ribosomes*.[‡] You can think of the mRNA as a tape on which the cell records information from a gene and sends it to the ribosomes for protein synthesis.

The cell controls protein synthesis by acting during RNA transcription. The cell exercises tight control over the genes so protein will be made when, and only when, the cell needs it.

The DNA also encodes the RNA the cell makes. Because proteins, and to a lesser extent RNA, dominate the cell's activities, the proteins a cell makes dictate to a large extent the nature of the cell. Because the DNA molecules encode protein and RNA, the information they contain, determine the nature of the cells. As a result, the DNA information determines the tissues, the organs, and ultimately the form of the organism itself.[§]

If the organism is to function properly, the right proteins must be made at the right time and in the right place. Each cell therefore has a control system that tells each of its genes when to turn ON and when to turn OFF. A gene is said to be ON when it's active, directing the synthesis of its protein; it is said to be OFF when it's not active.

---

[*] For how the amino acids are encoded in the DNA and RNA, see Appendix F.

[†] Appendix E gives some details of the transcription process.

[‡] Some details of protein synthesis in the cell are given in Appendix G.

[§] The organism takes shape under genetic control during development. The DNA alone does not fully determine the form of the organism. In Chapter 7 I shall point to the importance of external signals on the form of the organism, and the bearing they may have on large-scale evolution.

How does the cell control its genes? It turns them ON and OFF by means of special regulators. The regulators are most often proteins, and are known as *regulatory proteins.*[*]

Earlier in this chapter I told you that the embryo develops from a fertilized egg, and that this development follows a program much like a computer program. The genes make proteins that determine the state of the cell. The activity of a gene depends on the current state of its cell and the state of nearby cells. Nearby cells announce their state by sending out special messages in the form of protein molecules. The messages from other cells, as well as messages from within the cell, tell the gene when to turn ON and OFF. Each cell then develops by reading the incoming messages, generating outgoing messages, and following its genetic program,. The development of the entire embryo is coordinated through the messages each cell receives from other cells and from the outside, and through those it sends out to other cells.[†]

There are many different kinds of molecules that affect the activation of the genes. The largest known group of such molecules are the *hormones*. Each hormone is made in a special cell, which is part of a specialized gland. The glands that make the hormones constitute what is called the *endocrine system*. In an animal the glands discharge their hormones into the blood stream, which then delivers them to their target anywhere in the body. Hormones not only play a role in the control of genes, but they also can affect the action of enzymes and they can affect cell structures as well [Darnell et al. 1986]. They play an important role both in development and in normal body functioning.

---

[*] See Appendix H for how the cell controls its genes.
[†] A brief discussion of embryological development on the molecular level is given in Appendix I.

A cell divides to form two nearly identical cells in a process known as *mitosis*. Just before the cell divides, it duplicates its chromosomes forming a double set. During the division it gives one set of chromosomes to each of the two daughter cells. At every cell division each daughter cell gets a set of chromosomes identical with those in the mother cell.

Some cells of the embryo are not destined to become part of the body of the adult organism. The organism instead sets them aside to become the *germ cells*, which will make the next generation. The male parent makes male germ cells; the female parent makes female germ cells. The gametes are the daughter cells the germ cell makes when it divides in a process called *meiosis*. In meiosis the cell divides twice, but the chromosomes double only in the first division. During the second division the members of the chromosome pairs separate in what, as far as we know, is a random way. One member of each pair goes to each of the two daughter cells.

As a result each gamete gets one member of each chromosome pair. The gamete then has half as many chromosomes as the normal cell. The gametes are made by the parents. The chromosomes in the gametes turn out to be a random choice from those of the grandparents. Male and female gametes unite in fertilization to form the zygote.

When the chromosomes duplicate they don't do a perfect job of copying the DNA. They make about one error every 10,000 base pairs they copy [Darnell et al. 1986]. That's the error rate of a typist if he made one error every five pages. That error rate might be good enough for the office, but it's not good enough for genetic transcription. The genetic information has to be copied much more accurately than that to keep errors from building up over the generations.

To reduce the errors, the cell proofreads the DNA and corrects any errors it made in replication. But a few errors remain even after the proofreading. They are known as *copying errors*, or *single*-nucleotide *substitutions*. They are mutations belonging to a class known as *point mutations.** They are few enough for the species to tolerate. With the proofreading, the copying has a very low error rate from one in a billion to one in a hundred billion. One error per hundred billion would be like one error in fifty million pages of typescript. Fifty million pages are the lifetime output of about a hundred professional typists. That's some proofreading! The cell, or organism, can allow this small error.†

In addition to point mutations, there are genetic changes that can affect more than one gene. One such genetic change is known as *recombination,* in which two chromosomes, or parts of chromosomes exchange pieces. The exchange can take place as shown in Fig. 2.1. Part *a* of the figure shows a pair of chromosomes during meiotic division. Part *b* of the figure shows the two chromosomes crossing over each other. They both break at the crossing point. The chromosome pieces then switch, with the result shown in Part *c* of the figure. The two chromosomes of the pair have now interchanged pieces.

Recombination is not a simple process. We do not yet understand how the breaking of the chromosomes and the swapping of the pieces is done as precisely as it is. We do know, though, that it is controlled by special enzymes that

---

* In Chapter 1, I noted how the term *mutation* was introduced into evolutionary theory. Today we use the term to denote *any heritable change.* Any change in the genome of a cell that is inherited by the daughter cells qualifies as a mutation. A point mutation is one that affects only one nucleotide.

† Most evolutionists hold that these errors even play a *positive* role; they look on these errors as the source of the variation the neo-Darwinian theory needs. I disagree; I think these small errors just represent the limit of the accuracy with which DNA can be copied. Although I concede that they might play a role in small-scale evolution, I hold that they play no positive role in large-scale evolution. As we shall see in the following chapters, random variation cannot lead to large-scale evolution.

break the chromosomes, exchange the pieces, and rejoin the free ends.

> How such a double-stranded cleavage at two precisely analogous sites can take place, followed by the swapping of duplex regions and the rejoining of ends, is not at all clear. ... recombination is most often not a ... simple sequence of events [Darnell et al. 1986, p. 556].

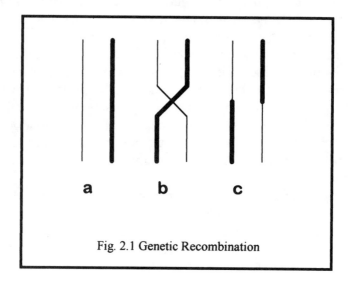

Fig. 2.1 Genetic Recombination

Not only is the process under enzymatic control, it also needs certain special structures in the cell to make it work [Darnell et al. 1986, Stahl 1987].

There are other kinds of mutations too. They are:

- *duplication* (or *amplification*) of a segment of the genome,

- *inversion* of a segment,

- *deletion* of a segment,

- *insertion* of a new segment,

- *transposition* of a segment from one place to another.

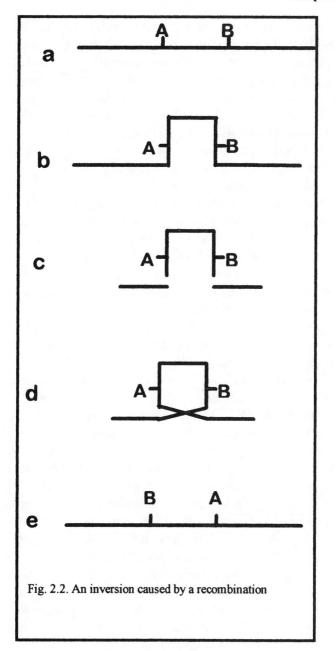

Fig. 2.2. An inversion caused by a recombination

An inversion can occur through a recombination within a single chromosome as shown in Fig. 2.2. In part *a* of Fig. 2.2, a genome is shown as a line with a segment marked off from *A* to *B*. This is the segment that will invert. A hairpin turn is formed as shown in part *b* of the figure. The segment *AB* then breaks at both *A* and *B* as shown in part *c*, and the broken ends of the segment are switched and rejoined to the broken ends of the main genome as shown in part *d*. When the genome is straightened as shown in part *e*, we see that the segment *AB* has been inverted.

Duplication of a gene can occur through a recombination where the crossover is not at the same point on both chromosomes of the pair, as shown in Fig. 2.3. Each chromosome initially has one copy of the gene. The upper one breaks at B and the lower one breaks at G. After recombination, the new chromosome ABGH will not have the gene, and chromosome EFGBCD will have two of them.

The segment between the place of the two breaks will then be duplicated in one chromosome and will be missing from the other.

Geneticists have found that the inversions, deletions, insertions, and transpositions are not just haphazard events. Special pieces of DNA that jump around in the chromosome cause these genetic changes. Short pieces of DNA, called *transposons*, have been found to jump from place to place in the chromosome. They have also been found to activate other special, shorter, pieces and make *them* jump as well. A piece of DNA that a transposon has activated is called an *insertion sequence* (*IS*). It is so called because it can be inserted into a gene. It is taken from one place in the genome and inserted at another. An IS, once inserted, can be deleted again. It can also be amplified into many copies, and it can even be turned around to make an inversion. Transposons and IS's can jump not only from one place to another on a chromosome. They can even jump from one chromosome to another [Darnell et al. 1986, Fedoroff 1984]. An IS even from a plasmid (a small

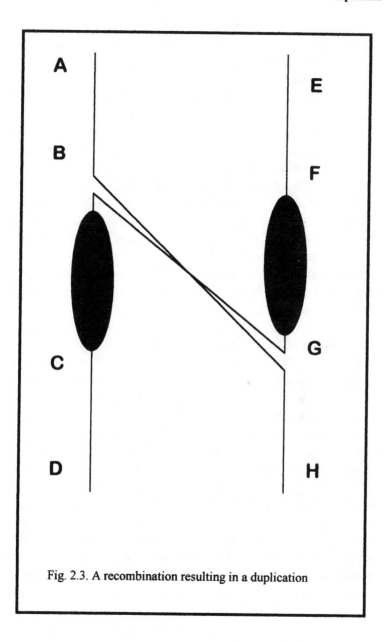

Fig. 2.3. A recombination resulting in a duplication

structure in the cell containing DNA that is not in a chromosome) can be inserted into the chromosome. Recombination between two identical IS's can delete the segment between them, duplicate it, or invert it [Stryer 1988]. In this way the IS's can be responsible for deletions, duplications, and inversions.

A transposon has in it sections of DNA that encode two of the enzymes it needs to carry out its job. The cell itself contributes the other necessary enzymes. The motion of these genetic elements about to produce the above mutations has been found to be a complex process and we probably haven't yet discovered all the complexity. But because no one knows why they occur, many geneticists have assumed they occur only by chance. I find it hard to believe that a process as precise and as well controlled as the transposition of genetic elements happens only by chance. Some scientists tend to call a mechanism *random* before we learn what it really does. If the source of the variation for evolution were point mutations, we could say the variation is random. But if the source of the variation is the complex process of transposition, then there is no justification for saying that evolution is based on random events. I'll return to this subject in Chapter 5 and again in Chapter 7.

After a gene mutates, the cell preserves the new gene with the same care it gives to any other gene. It copies the change in later cell divisions with the same fidelity as it does the rest of the chromosome.

A mutation in a gene can change the effect of the gene on the organism. An error in one base of a codon[*] can change the amino acid it encodes. The amino acid will be changed unless both the old and the new codons happen to encode the same amino acid. A change of an amino acid in a protein of-

---

[*] See Appendix F for a definition of *codon*.

ten results in a change in the properties of that protein.* A change in some amino acids may make only minor changes in the protein's performance. But in others it can make major changes.

There are several examples of a single nucleotide substitution known to have potent effects on humans. One is the substitution of a T for an A in the gene encoding one of the chains in the human hemoglobin protein. This change is in the codon for the sixth amino acid of what is known as the β chain of the protein. The mutation changes a glutamic acid to a valine and results in a serious disease known as *sickle-cell anemia* [Luzzatto and Goodfellow 1989].

Another example is the change from a C to an A in the gene that encodes a pigment, called *rhodopsin*, in the retina of the eye. As a result, the 23rd amino acid of the rhodopsin protein changes from proline to histidine. This single change results in an eye disease that leads to blindness [Dryja et al. 1990, Applebury 1990].

A mutation of a single nucleotide in a regulatory gene can change the way it regulates. A mutation in a regulatory gene translates into a change in the regulatory protein it encodes. The change could be in the strength with which the protein binds to the gene it controls. RNA polymerase is an example of a regulatory protein; a repressor is another. Sometimes a change of a single amino acid destroys the protein's ability to function altogether.

A mutation in a gene encoding an enzyme often leads to a change in the way the enzyme functions. A mutation in some regulatory genes can change the way an embryo develops. If the mutation is in a germ cell, the change will be heritable and will appear in all later generations. A change in the phenotype can be subject to the force of natural selection.

---

* Observations of the function of a protein outside the living cell often do not tell the whole story. Some changes in a protein may be evident only in a living cell.

Recall that one of the two parts of the neo-Darwinian process is random variation in the genotype. According to many evolutionists, the major source of this variation is the copying error. Other kinds of genetic change, such as those listed above, also can occur. But Darwinian theory asks that the mutations be both spontaneous and random. If those in my list are indeed spontaneous and random they can't do better for evolution than can the copying errors. The copying errors have more flexibility in making new sequences than any of the other mutations have.[*]

———

Each gene in a population can come in more than one version, and each version of a gene is known as an *allele*. Each allele can have its own effect on the phenotype. A red and a white flower, for example, are the expressions of two alleles of the same gene. So are blue and brown eyes in humans. There can be many alleles of the same gene in a population, and a single organism will often have two alleles of the same gene, one in each of the two members of a chromosome pair. The set of alleles in the genome of one individual almost always differs from those of another. These differences are thought to have come from mutations that have occurred in the past.

One member of each pair of chromosomes comes from the father and the other comes from the mother. The two members of a pair have the same genes, but their genes are c .ten different alleles.

———

[*]An arbitrary sequence of base pairs can be made more easily through a sequence of point mutations than it can be made through sequences of inversions, deletions, or additions. In general, one arbitrary sequence cannot be converted to another through a sequence of inversions alone. A trivial example is the conversion of poly-A (AA...A) to poly-G (GG...G). This clearly cannot be done by any sequence of inversions, but it can be done by the proper sequence of point mutations. Deletions of segments of base pairs followed by insertions of random segments can change one arbitrary sequence into another, but clearly, it can be done in far less steps by point mutations.

If each of two parents differ by only one allele in each chromosome, there will be a large potential variety in their offspring. The human, for example, has 23 pairs of chromosomes. The mother can make female gametes drawn from a set of more than 2^{23}, or 8,400,000 different ones.* The father has the same potential in making male gametes. The offspring that could come from the two of them together can then be drawn from a set of more than 8,400,000 × 8,400,000, or about 70 trillion. That's about ten thousand times the population of the earth. The potential for variety is large.

But recombination can multiply that variety greatly. As a germ cell divides in meiosis and the chromosomes swap the pieces between them, many combinations can appear. If each chromosome could break in only one place, then instead of shuffling 23 chromosomes from each parent, meiosis could shuffle 46 pieces of chromosomes from each parent. If the members of each pair of chromosomes differed in both these pieces then the number of possibilities in a single gamete would then be 2^{46}, or about 70 trillion. Between the two parents, the number of combinations climbs to almost 5×10^{27}. Actually the chromosomes can break in more than one place, so the number of combinations can be much larger.

———————

---

* For each of the 23 chromosome pairs, a gamete can get either member of the pair. There are 2 ways of picking a member for the first pair. There are 2×2, or 4, ways of picking a member each for the first and second pairs. The number of ways of picking a member each for all 23 pairs is the product of 23 two's, or 2^{23}, or about 8.4 million. I am here assuming the mother has, in every chromosome pair, two different alleles for at least one gene, and similarly for the father.

NOTES TO CHAPTER 2

1. The nucleotides of the DNA chain. The links in the DNA chain are called *nucleotides*. A nucleotide is made of what is known as a *nucleoside*, with an attached phosphate group. The phosphate group is an atom of phosphorus joined with four atoms of oxygen, and it is the link joining the nucleosides. It joins them by sharing one of its oxygen atoms with each. A nucleoside is a combination of two molecules; one is a sugar; the other is called a *base*. In DNA, the sugar is *deoxyribose* (that's where DNA got its name). The deoxyribose sugar is a molecule of 19 atoms, and the bases are molecules built from between 12 and 16 atoms each, depending on the base.

Chapter 3

THE NEO-DARWINIAN THEORY OF EVOLUTION

MY cousin Dorothy is an elderly lady, much older than I. She has several great grandchildren. She once told me that when she was a little girl she used to have long conversations with her great grandmother. Her great grandmother used to tell her stories of what she remembered when she was a little girl in Russia when Napoleon invaded in 1812. I find it exciting that Dorothy heard some history of the early nineteenth century first hand, and has brought those stories forward to the 1990's, more than a hundred and eighty years later! Moreover, her great grandchildren hear those stories only second hand, and they can carry them forward well into the second half of the twenty-first century.

Most of us have a deep interest in who our ancestors were and what they did. Some lucky ones hear of traditions in their families that go back even more than two and a half centuries. But most of us have to be satisfied with reading history books and imagining how our ancestors lived and what they did. We're all interested in our roots.

Who was my great, ..., great grandmother? How far back can we go? The further back we go the hazier our information gets. Do we all trace back to Adam and Eve? Or maybe we need only go back to Noah and Naamah. That was the conventional wisdom of the Western World until 1859. Now that world is divided. Most scientists (not just biologists), and many others too, believe that we go much further back than that. Somewhere back in the deep past, they say, our great, ... great grandfather was an ape-like creature, covered with

49

thick hair instead of clothing. Further back, they say, our ancestor was some kind of insect-eating mammal. Still further back in our family tree our ancestors were fish.

Was my great, …, great grandmother really a fish? Was her great … great grandmother something like a slime mold? Did we come to be what we are just through the natural sifting of random events? That's what the neo-Darwinian theory tells us, but I don't find the theory convincing. Let's first see what the theory says, and then we'll analyze it.

As described in Chapter 1, Darwin's original theory was revised in the 1940's into the *neo-Darwinian theory (NDT)*. It was established to answer the difficulties that had arisen with new discoveries in biology. The NDT was perceived as answering the problems with evolutionary theory, and it has served as the most widely accepted theory for more than 50 years.

But science does not stand still. Discoveries in the last fifty years, and particularly in the last twenty, have been forcing evolutionists to patch in changes to the NDT. With the natural conservatism of good science, they have been trying to keep those changes minor. The changes nevertheless are stretching the bond that holds the theory to the facts. That bond has been stretched to the breaking point, and evolutionists will soon have to acknowledge that the bond has already snapped. They shall soon have to make a major innovation that will change the theory fundamentally, and force a basic change in the philosophy of evolution. A new theory of evolution is desperately needed.

———

The NDT is about selection giving direction to variation created by an underlying random process. The evolutionists have succeeded in treating selection in a quantitative way. Indeed they have established the field of *population genetics* to do just that. They have also been able to point to a source for the random variation which plays a prominent role in the theory. In the NDT they have retained the two basic Darwin-

ian novella of *variation* and *natural selection*. They have,
however, sharpened Darwin's sometimes confused notion of
variation, and insisted that it is *spontaneous and random*.

Offspring are like their parents because the process of heredity is stable. If it weren't stable we wouldn't recognize
chickens as chickens nor dogs as dogs. But the stability isn't
absolute: a child is not an exact copy of his parent.

Even if a population started from a single male and female
pair, it wouldn't stay uniform. Variation would arise from
mutations in the germ cells of the first parents as well as in
those of later offspring. More variety would come from recombination and the shuffling of the alleles already in the
first parents. The differences among those alleles are thought
to have arisen from mutations that occurred in past generations. All populations have variety in them, and more of it
continues to appear.

A basic dogma of the NDT is that mutations are not at all
related to the needs of the organism. The theory is based on
random genetic changes. If evolution were found to be driven
by genetic changes that were not random, that were in any
way a response to the environment or to the needs of the organism, it would contradict the theory. Evolutionists have
stressed time and again that the variations from which evolution stems are *random* in this sense.

———

Heritable variations fall into three classes according to
whether their effects are positive, negative, or neutral.

1. *Positive* variations are those that help an organism
   have more surviving offspring.[*] The variation can affect the number of surviving offspring in both direct
   and indirect ways. An increase in fertility is an example of a *direct* effect. A variation that helps an
   animal run faster is an example of an *indirect* effect.

---

[*] In this sense, an organism is said to "survive" if, and only if, it lives
to reproduce.

A faster animal can get food more easily. It will have more time and energy to mate, and it will have a better chance to reproduce and will tend to give birth more often. A faster animal can also better escape a predator, and as a result it will tend to live longer. Its longer life will permit it to have more offspring.

2. *Negative* variations are the opposite: they reduce the number of surviving offspring. They might affect fertility *directly*, but they might act in more subtle ways as well. For example, the individual may not see so well and may not be able to gather food so easily as the others. It also won't be able to evade predators so well, and as a result it won't live so long as the others. An extreme example of a negative variation is one that kills the organism before it can reproduce.

3. *Neutral variations* have no effect on fertility and they make no difference to evolution. They do not influence the organism's ability to survive to have offspring. Eye color in humans, for example, seems to be a neutral variation.

Selection must be thought of as acting not on an individual but on a population. If an organism had a mutation that we call positive, it will tend to have more offspring than the average of the population. If it had a negative mutation, it will tend to have fewer offspring. The NDT then says that the number of those having positive mutations will tend to increase and they will eventually take over the population.

If there were no limitations on a population, it would increase forever. One animal, for example, might have 10 offspring. If they all live and reproduce, each of them might have 10 offspring, making 100 in the third generation. If the birth rate continues at the same rate there would be 1000 in the fourth generation, and so on. By the seventh generation there would be a million, and by the tenth there would be a billion. For growth to continue at the same rate to the twentieth generation the bodies alone of that large population would need nearly all the nitrogen and carbon there is on earth!* This type of rapid increase, where the number increases by the same factor in each generation, is known as *exponential growth*.

The growth is exponential only as long as the population has not yet reached any of its limits. The population cannot continue to grow forever at the same rate because it will at some point reach a limit, such as in its food supply or living space. When that happens the death rate rises or the birth rate falls.†

Eventually the population will reach an equilibrium when the death rate equals the birth rate. At equilibrium the population is constant: it neither increases nor decreases. Outside restraints limit the size of the population. When the population is at its limit, one part of it can grow only at the expense of the rest.

Those individuals that have more offspring can be called more *fit* or more *adaptive*. The theory assumes that a single mutation can make an organism more adaptive. Because the offspring of the more adaptive organisms inherit the parents' adaptive traits they also will be more adaptive than the norm and will tend to reproduce more. The number of descendants of the adaptive mutant will then tend to increase as time goes on. At the start, their increase will be exponential. In the end, so the theory goes, the adaptive group will take over the population.

---

* That's assuming each animal weighs about 10 kilograms. There are about 10^{17} tons of carbon in the earth's lithosphere and about 10^{16} tons of nitrogen (including what is in the atmosphere) [Needham 1965]. In the 20th generation there would be 10^{19} animals, having a total body weight of 10^{17} tons.
† Animal populations are known to control their own birth rate so as not to outstrip their food supply [Wynne-Edwards 1986]. As a population approaches its natural limit, its birth rate falls.

For an organism to play a role in evolution it has to reproduce. Organisms that don't reproduce don't leave their genetic mark in the population.

I can quantify the "fitness" of a phenotype by assigning a number to what I shall call its *selective value* (*SV*). I define the *SV* of a phenotype as the fraction by which its average number of surviving offspring exceeds that of the population norm. For example, a mutant whose average number of surviving offspring is 0.1% higher than the rest of the population has an *SV* of 0.1% (or 0.001). A *negative SV* means that its survival rate is *less* than that of the rest of the population. A mutant with an *SV* of *minus* 0.1% will have a survival rate 0.1% *less* than the population norm. The *SV*'s of *positive* mutations are *positive*. The *SV*'s of *negative* ones are *negative*. Natural selection will tend to increase the number of individuals with a positive *SV* and decrease the number with a negative *SV*.

Sir Ronald Fisher was a mathematician and one of the world's experts on the mathematics of evolution. He was one of the architects of the NDT and one of the founders of the field of population genetics. He made one of the first mathematical studies of how natural selection works.

Fisher has shown that most mutants, even if they have positive *SV*'s, will be wiped out by random effects. He noted that a single mutation, even if it is a positive one, has only a small chance of survival.[*1] As a result, a single mutation is unlikely to play much of a role in evolution. Fisher concluded that if positive mutations are to play a role in evolution, many of them have to occur [Fisher 1958].

The late George Gaylord Simpson, was a well-known paleontologist and a leading spokesman for evolution. He acknowledged that a single mutation has little chance of staying in a population. But he thought that positive mutations would

---

[*] If your are wondering how random effects can wipe out a mutation even if it is advantageous, you will find a simplified and intuitive explanation in Note 1 at the end of the chapter.

occur often enough that they "may readily become established in populations" [Simpson 1953, p. 118]. A careful examination of the subject shows that his estimate is wrong. We shall examine the matter in the next chapter.

I used Fisher's method to compute the chance that a gene will survive. I found that in a large population, a genome with an *SV* of 0.1% will have only one chance in about 500 of surviving the random effects that tend to wipe it out. If there were 500 mutations with an *SV* of 0.1% the chance that at least one would survive would still not be 100%. It would be only about 5 out of 8.* If there were 1000 of them, their chances would be about 6 out of 7. Only when there would be about 2500 would their chance of surviving be more than 99%.[2] So positive mutations will take over the population only if many of them occur.† Otherwise there's a good chance they will disappear.

Fisher's result contradicts Darwin's intuitive notion of natural selection preserving even the slightest variation and increasing its numbers until it takes over the population. Darwin erroneously thought that even the smallest improvements would be selected. He said:

> ... slight modifications, which in any way favoured the individuals of any species, by better adapting them to their altered conditions, would tend to be preserved. ... Under nature, the slightest differences of structure or constitution may well turn the nicely balanced scale in the struggle for life, and be so preserved. ... Can we wonder then that Nature's productions should be far "truer" in character than man's productions; that they should be infinitely better adapted to the most complex conditions of life, and should plainly bear the stamp of far higher workmanship? It may metaphorically be said that natural selection is daily and

---

* It's more nearly 1-1/e
† The 500, 1000, or 2500 mutations I am discussing here need not be identical, of course. The chances of survival I cited hold if the mutations all have about the same *SV*. Evolution is served if any one of them survives to take over the population.

hourly scrutinising, throughout the world, the slightest variations; rejecting those that are bad, preserving and adding up all that are good; silently and insensibly work-ing, whenever and wherever opportunity offers, at the im-provement of each organic being in relation to its organic and inorganic conditions of life. [Darwin 1872, pp. 82-84].

Fisher's mathematical analysis of natural selection shows that Darwin was mistaken. Amazingly, the neo-Darwinians, who included Fisher, preserved Darwin's error in their theory. Fisher thought that adaptive variations were highly probable.

When neo-Darwinian evolutionists write for students or for the public they often gloss over the point Fisher made. For example, Theodosius Dobzhansky, who was a highly re-spected geneticist and a leading evolutionist, wrote in his widely-used text book:

If the carriers of one genotype produce on the average 1,000 offspring when the carriers of another genotype pro-duce 999, the difference in the adaptive values will in time bring about a change in the genetic composition of the population [Dobzhansky 1951, p. 79].

What he said holds only if there were many of the first genotype at the outset. If there was only one mutation, the chances can be 500 to 1 that it would disappear before it could influence the population's genetic structure.

Small populations promote the survival of a single gene more than large ones do, but they pose a dilemma for evolu-tion. Indeed, a lone positive mutation has a better chance of surviving in a small population than in a large one.[3] It also has a better chance of taking over a small population.

Consider, for example, a population of 100,000 butterflies. A single mutant, even with a positive SV, in this population will have only a small chance of surviving among the 99,999 others. But suppose ten of these butterflies are blown by a hurricane out to an island that has no butterflies and from which they cannot return. A mutant with a positive SV on the island would have a better chance of taking over its population of 10 than would a mutant on the mainland have of taking over its population of 100,000.

But a population of ten can more easily be wiped out by a chance disaster, like a drought or a tidal wave, than can 100,000 of them. The mutant's fitness wouldn't help here. There's also a much smaller chance that a positive mutation will even show up among the ten than among the 100,000. On balance it turns out that the chance of a positive mutation occurring and staying will be larger in larger populations.[4]

Darwin saw nature doing what a pigeon breeder does. The number of organisms that reproduce better would grow as if they were being selected. That's why he called the growth of their numbers *natural selection*. Those who study natural selection use terms like *selection pressure* and *the force of natural selection* to refer to the speed with which a mutation spreads through the population.

Evolutionists claim that over long enough times there can be a large change in a population. A small mutation might occur in a germ cell and produce an individual that differs a little from the norm. If the mutation is negative it will have a negative SV and will tend to disappear. If it's positive it will have a positive SV.

If enough mutations with positive SV occur the number of the new type will tend to increase. In time it will take over the population. After one change has been fixed in the population another mutation may occur and the process repeats. One small change adds to another and the process goes on for a long time. The NDT says that large changes will eventually result. It's like becoming a millionaire by saving enough pennies.

———

When the neo-Darwinians set up the NDT, no one knew the molecular nature of the variation such a theory would require. When Watson and Crick reported their discovery of the physical and chemical structure of DNA in 1953, they suggested that errors in single nucleotides could occur during replication. Such errors were indeed found, and they became known as *copying errors*. The neo-Darwinians identified those errors with the random variation needed by the NDT.

The copying errors seemed to the neo-Darwinians to fill the role well. As far as we know, they are indeed random. To say they are random, though, is not to say that they can have no known cause. Although some seem to be just mistakes and do not have any known cause, higher rates of mutation can be induced by chemicals (called *mutagens*), or by ultraviolet (UV) radiation.* Nevertheless, these mutations still deserve to be called random. Their physical cause dictates only the frequency of the mutations, and not their effect on the organism. Thus bombardment of DNA with UV photons will increase the chance that the DNA molecule will change. But it will not make it change in a way that, for example, makes the organism resistant to UV.

There is another characteristic of the copying errors making them suitable, in neo-Darwinian eyes, for the NDT: they are small. The NDT had to rely on small mutations to answer some of the objections that had been raised to Darwin's theory.

––––––––

You can compare mutations in the nucleotides of DNA to making random changes in the letters of a book. Suppose an author tried to improve his novel by changing letters in it at random. If he did it the way evolution is supposed to work he would reprint the novel with a few random mutations of the letters. Then he would read it and see if he liked the changes. If he followed the rules of evolution he would have to pick

––––––––––––––––––

* The UV radiation itself does not directly cause the mutation. The UV photon is absorbed by a DNA base molecule, boosting it into an unstable high-energy state. As the molecule drops back to a lower-energy state it deforms. A cellular-repair system, called *SOS*, is then activated to delete the deformed bases and replace them with good ones. The SOS system is, however, inaccurate and it often substitutes the wrong base.

either the altered text or the original one. He could not choose among the individual changes. He must take the whole package or nothing.

As in neo-Darwinian evolution he would have to make the mutation rate low. He could not tolerate a change of much more than one letter in the novel at each trial. If the mutation rate were too high he would get many changes, and most of them would be negative. Every time he would get an improvement he would be likely to get an injury with it, and he would have to reject it.

Suppose an author got a mutation of one letter in his novel. He might, for example, get the word *face* to change to any of the words:

dace, fact, fade,
fake, fame, fare,
fate, faze, lace,
mace, pace, race.

Or he could get a change to any of 88 other combinations of four letters that are not words. Let's take a sentence from Charles Dickens's *Tale of Two Cities*:

It portended that there was one stone face too many up at the château.

Suppose the word *face* in this sentence mutated to *fake*, and suppose Dickens liked the new meaning better. He might then prefer it to what he had before. But a change in meaning in one place often needs other changes elsewhere. Even though the change to *fake* makes potential sense, it doesn't fit with anything that comes before or after it. If Dickens wanted to avoid the nonsense of changing the word *face* here to *fake*, and exploit its potential for new meaning, he would need other changes. If Dickens wanted to write about a *stone fake* he would have to build up to it, and then he would have to follow it through. Furthermore, he would have to remove any earlier mention of a *stone face* if he had any. Without these other changes the text would be better if he left it as it was.

You can see here one of the problems with Darwin's theory and the essence of what was one of the early objections to it. The objection was that, to get an improvement in a species, several correlated adaptations have to occur together. These requirements are far too unlikely for the theory to work.

Both Darwin, and the neo-Darwinians after him, tried to solve the problem by saying that mutations are small, and they make only small changes in the phenotype. They say small changes will occur first in one place in the genome and then in another.

They'd like you to think of evolution as something like climbing up the inside of a well. Your left foot pressed against the wall will not hold you unless your right foot is pressing opposite it. If you try to climb, you'll start to fall as soon as you release one foot and try to raise it. But if you release the pressure on your feet for just an instant, enough to move up a fraction of an inch, you can stop yourself before you fall. You can climb this way with small steps if you always keep your feet nearly opposite each other on about the same level. The neo-Darwinians think evolution could work something like this. They said it will work if the changes are small enough.

Because the copying error is the smallest possible mutation, they were happy to adopt it as the source of the variation for their theory. But a copying error is not infinitesimally small. In my example of the novel I made the smallest change possible by changing a single letter. I can't change *less* than one letter. The same is true for living organisms — they can't change by less than one nucleotide.

Evolutionists have given only a vague answer to the objection of correlated adaptation. They haven't shown that the solution of the small change, which Darwin used to explain how the eye could have evolved, will indeed work. Evolutionists have not shown they have a solution that will let the NDT work.

The NDT says a population will tend to change to adapt to its environment. It says changes appear at random and build up. Natural selection, in a way, looks at each change and decides to drop it or keep it. The NDT says this is how adaptive changes build up over a long time. Many hold that, as populations have adapted over long periods of time, their members have become more complex. This is how the NDT tries to explain how the complexity of life built up. Richard Dawkins has said:

> The one thing that makes evolution such a neat theory is that it explains how organized complexity can arise out of primeval simplicity [Dawkins 1986, p. 316].

But as we shall see in Chapter 5, there is no evidence that complexity has been built up by the process described by the NDT.

———

Only a *heritable* change can play a role in the NDT. A change that is not heritable can have no effect in evolution. If a wolf loses an eye or breaks a leg, for example, it will not pass on the change to its descendants.

The evolution of single cells is best achieved through mutations in the genes that encode enzymes. That's how bacteria are supposed to have evolved. But multicellular organisms would need other kinds of changes as well. For a fish to evolve to a frog, there would have to be changes in the genes that control development. How else could a fish evolve into a frog, or a dinosaur into a bird?

———

What is the source of the genetic variety in a population? Its ultimate source is mutations. Copying errors are rare, and they can build variety only slowly. Evolution that feeds on this variety must be slow — slow enough to cause serious problems for the NDT.

Some time ago I looked at how the rate of copying errors affects the rate of evolution [Spetner 1964]. It turns out that the mutation rate is limited by the need to maintain the integrity of the species. If the mutation rate is too high, too many individuals would have one or more of their genes damaged. Because experiments have shown that most mutations are harmful, genes that had already built up to be useful would suffer damage.[5] The mutation rate has to be low, making the rate of evolution slow — once again, too slow for evolution to work the way the NDT says it does.

Mutation is the source of the variety the theory calls for. Genetic recombination also can be a source of variety. In fact some evolutionists hold that the main and immediate source of variety is genetic recombination. But ultimately the source must be mutation. Only mutation has the flexibility to change the genome so it will generate new structures and functions. Sir Julian Huxley, one of the group who set up the NDT, pointed out:

> Mutation is the only begetter of intrinsic change in the separate units of the hereditary constitution: it alters the nature of the gene. Recombination, on the other hand, though it may produce quite new combinations with quite new effects on character, only juggles with existing genes. [Huxley 1943, p. 21].

New DNA sequences come only through mutations. Recombination can't do much more than bring out what's already there.

Evolutionists have suggested that a population could store mutations for later use. They say that even if the mutations aren't useful now, maybe they can sit in the population until they become useful. They expect this kind of storage to make up for mutations being slow and rare. As we shall see in Chapter 4, the chance of getting a positive mutation just when it's needed is very small. Michael Ruse, for example, speculated:

> ... that there are usually masses of variation held in any natural population, and that selection can get straight to work whenever the case arises. It is not necessary to wait for the appropriate new mutation [Ruse 1982, p. 94].

Some evolutionists have proposed another way to store mutations for later use. A gene can have a piece of it turned around (from an inversion), or it could have an extra piece of DNA (from an insertion) that would keep it from working. The latter could shift the coding frame.* Then when the right inversion or deletion occurs it will correct the coding, and the gene will suddenly be functional.

According to this proposal the gene is stored like a piece of knocked-down furniture, waiting only for the one inversion or deletion that will make it work. When that mutation appears, the gene can function.

There's a lot of room for variability in DNA. You could measure it by counting all the possible DNA sequences. For mammals, for example, there are about $10^{24,082,400}$ such sequences. I'll call this number the *variability* number.[6]

If you wanted to write this number you'd have to write a one followed by more than 24 million zeros. If you wrote 2000 zeros to a page, you'd need 12,000 pages to write the number. If you bound those pages into books, the books would take up about a foot and a half of shelf space and that's just to write the number! The variability number of an organism is a number so large we don't have a feel for its size. And such a number represents the potential genome variety.

But what about the genome variety that's actually in a population? Does it at all reflect the variability number of the

---

* As explained in Appendix F, codons, which encode amino acids, are triplet strings of base pairs. If a gene had one extra base pair, the coding frame (the way the triplets are grouped) would shift by one base pair. From that point on the code would be nonsense.

genome? What fraction of the $10^{24,082,400}$ possible genomes could be in a real population? Contrast the huge number of possible sequences of the DNA with the size of an animal population. A population might have a hundred thousand members, or a million if the animals are small. Suppose the population is even a billion. The fraction of all sequences that could be stored in the population would then be a billion divided by $10^{24,082,400}$, or $1/10^{24,082,391}$. This is the fraction of the DNA sequences that can be stored directly in the population. The number is so close to zero it's negligible. The genome variety found in a population is in no way a match for the potential variety of the genome.

Some say that genes that could be adaptive in a future environment might be in two or more separate pieces. They wouldn't be active in this form, but maybe a recombination event could activate them. There are many more ways to put the pieces together than there are members of the population.

Francisco Ayala of the University of California has made an estimate of how much variety a human population could store [Ayala 1978]. There are some 100,000 genes in the human genome. Of these, about 6,700 come in two versions in the same person, one on each member of a chromosome pair. There are then $2^{6,700}$, or about $10^{2,017}$, different possible combinations. That huge number is far bigger than any population. In fact, it's very much larger than the number of all the protons in the universe. As far as anyone knows, a meiotic division of a cell could result in one of those many combinations.[*]

But so long as these stored genes are not active, there's no selection. If there's no selection, these genes won't be treated any differently from any other inactive gene that has no selective value. Without selection, all strings of DNA have the same right to be in the genome. How can a population know

---

[*] But we don't really know enough about the mechanism of recombination to say with any certainty that *any* possible combination can occur.

to store a DNA string that will some day be useful? Only by brute force! It will have to store with it a massive number of other DNA strings that will never be useful. Ayala's number of $10^{2,017}$, as big as it is, is insignificantly small compared to the $10^{24,082,400}$ possible combinations in the DNA. The chance would be very small that any of those recombinations could make an improvement in the gene.*

There are some alleles in populations that have proved to be useful in the past. But you can't expect to find new genes lying around that can play a useful role they never played before. Alleles that have been useful in the past do get stored in a population, and are found in large numbers, closer to thousands than to 1. That may be why, when useful genes are found, there are so many of them. They might be left over from what was once a large fraction of the population, and maybe selection pressure has not yet got rid of them.

Genes that were once useful but aren't now, could still sit in the population. The more there are, the longer they can stay dormant in the population. Some genes would be adaptive now if they could get put together right. They may need only a recombination or an inversion to reawaken them. Others could be in the population in working order, but would not be adaptive now. They could lurk there, hardly noticed, until they are needed once more. When they are needed they would be selected, and their numbers would grow. Such examples would not demonstrate the effect of random variation.

Alcohol resistance in vinegar flies is an example of an allele that lurks in the population in working order. The gene for it exists in functional form in a small part of a population [McDonald, et al. 1977]. Some resistant flies are in the population all the time. Their numbers increase when conditions select for them.

---

* If you should assume there are enough adaptive combinations so that even from this small fraction you would have one, then you get into another problem that keeps evolution from working. I discuss this point in Chapter 4.

Another example is the peppered moth, which Kettlewell first reported in the 1950's [Kettlewell 1955, 1959, 1973]. Since then many evolutionists have cited it as a good example of Darwinian evolution in action. The peppered moth has changed from a light to a dark color and then back to a light color, and it seems to have done so through natural selection. Fig. 3.1 shows the light and dark forms of the peppered moth.

The peppered moths live in Great Britain. Before the coming of industry with its pollution they were of a light color peppered with small dark spots. Normally, the moths spend their days on lichen-covered tree trunks. The moth's spotted light color serves as a camouflage against the background of the light-colored lichens and protect it from birds that prey on it. The rise of industry in Britain spread soot in the industrial areas, and the soot blackened the buildings and trees. On the soot-blackened background the moth's light color was no longer a camouflage. But within about 100 years (or less) the moth population in industrial areas evolved a protective dark color. In the rural areas, however, most of the moths remained light.

Melanin is the protein that gives the dark color to the skin and hair of animals. The skin cells of dark moths make more melanin than the light moths do, and that's what makes them dark.

Pollution control in the industrial areas was started in the 1960's and has since cleaned up the environment. After the air was cleaned of soot, the lichen-covered tree trunks regained their light color. The moths have also, to a large extent, returned to their former light color [Cook et al. 1986].

Chapter 3

Evolutionists often cite industrial melanism as an example of evolution. Although it may be an example of natural selection, it is not an example of random variation. It turns out that when the soot began to cover the lichens, the light-moth population didn't have to wait for a mutation to turn dark. The dark moth was already in the population. It was living as a small minority among the light moths [Bishop and Cook 1975]. Where the tree trunks are light, most of the moths are light colored. Where the tree trunks are sooty, most of the moths are dark. There was no *random* variation. Both types of moths have been living side by side in both environments.

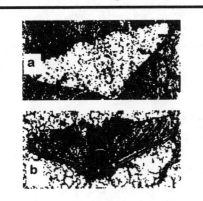

Fig. 3.1. (a) Light-colored peppered moth on dark background. (b) Dark-colored moth on light background. (Adapted from Bishop & Cook [1975] by permission)

There are two terms that I shall use often. They are *microevolution* and *macroevolution*. *Micro*evolution is a small evolutionary change. It can sometimes lead to a new variety or a new species. *Macro*evolution is supposed to be a large change in a population. A macroevolutionary change can lead from one major group to another, such as from a fish to an amphibian, or from a reptile to a bird. The attraction of *The Origin* was its promise to account for *macroevolution*.

Kettlewell's example of the evolution of the peppered moth is microevolution. So is the evolution of alcohol resistance in vinegar flies. So is the change in the shape of snail shells.

67

Macroevolution, on the other hand, would be the change from a dinosaur to a bird, from a fish to a frog.

Darwin, and the neo-Darwinians after him, explained macroevolution as the result of many steps of microevolution. A small step, they explained, is the result of a small variation acted upon by natural selection. They needed no further mechanism to explain the large changes. They were just the gradual accumulation of many small ones.

In recent years, some biologists have pointed out what they hold to be a serious problem with the gradualness of the NDT. They find fault with the thesis that macroevolution is made up of long chains of many small steps. Some have suggested that the gaps in the fossil record are real, that there really were abrupt changes in animals in the past. They claim that most evolution is made up of large abrupt changes [Eldredge and Gould 1972, Gould and Eldredge 1977]. They hold that populations remain in unchanged equilibrium for long spans of time that are punctuated by large abrupt changes. They have called their theory *Punctuated Equilibrium*. They hold that macroevolution is not just an accumulation of many small steps. When they first proposed their theory they held that large changes are qualitatively different from the small ones. Macroevolution consists of large abrupt changes [Gould 1980, Stanley 1979].

More recently, some punctuationists have changed their minds. They have returned to the neo-Darwinian fold and hold that macroevolution occurs not through jumps, but through cumulative selection. But they differ from the neo-Darwinians in that they hold that macroevolution could occur in smaller populations and in fewer steps than had been previously thought [Eldredge and Gould 1988, Maynard Smith 1988]. The continuous changes only *appear* to be jumps in the fossil record.

For convenience let me use the name *brady* for one who holds that evolution goes slowly and gradually. I would include in this category not only the neo-Darwinians, but the reformed punctuationists as well. One who holds evolution to proceed in leaps and jumps, through single large mutations, I shall call a *tachy*. These include the unreformed punctuationists.

The tachys not only claim the fossil record as support for their view, they also claim support from the way the genes of species differ from each other. In each step of evolution there must be a change of at least one nucleotide. If major changes come only through long chains of small steps, then in each major change in phenotype, there must be a change of many nucleotides in the genome. Organisms belonging to different major groups would have to differ by many nucleotides.

We find some groups of very different mammals that do not differ much in their protein or DNA sequences. Yet there are two very similar frog species that are held to have been separate for an estimated 150 million years whose proteins have many differences [McDonald 1990, Wilson et al. 1974]. The frogs show that the accumulation of many copying errors does not have to lead to a large difference in the phenotype. The mammals show that major differences can come from just a few changes, perhaps in regulatory genes.

The bradys do not see these problems as serious enough to abandon cumulative selection. But to the tachys, they are serious. The tachys want to explain how evolution might go by leaps and jumps. Some tachys have suggested that a large change comes about not through a long chain of copying errors but through a single complex genetic rearrangement [McDonald 1990, Stanley 1979]. Both bradys and tachys agree, however, that whatever the mutations are, they are random. That means they are not related to the needs of the organism and are not biased toward adapting the organism to its environment.

You will recall that genetic rearrangements include *inversions, insertions, deletions, transpositions*, and *amplifications*. Examples of each of these types have been shown to result from the action of mobile genetic elements, or transposons. Such rearrangements can change the way the gene performs. If any of these mutations occur in a germ cell, the new gene structure will be heritable.

The tachys look for the effects of these kinds of mutations in regulatory genes. The insertions, transpositions, and inversions, turn OFF the gene in which they occur. A deletion, if it precisely deletes a previous insertion, will turn the gene back ON. An inversion can also turn a gene back ON if that gene were turned OFF by an inversion, and if the second inversion exactly undoes the first.

———

Any theory meant to explain how all plants and animals evolved from a single cell has to explain how all the information got into the genome. All living organisms have a lot of information in them. You can gain some idea of how much information is in an organism both from the size of its genome and from the complexity of both the organism's structure and its function.* Information and complexity go hand in hand.

According to the doctrine of evolution, all the information in life today was built through evolution. If the neo-Darwinians think their mechanism can explain how evolution took place, they have to show how that mechanism could have put large amounts of information into the genome. They have to explain how the genomes got all the information they have.

According to the bradys, the information in the genome builds up slowly over a long time through long chains of small evolutionary steps. Each small step has to do its share of the job of adding information to the genome. Because each step is small and can add no more than a little information,

———

* As I have already noted in Chapter 2, the amount of DNA in a genome is not a good measure of the information in it. Some genomes contain a lot of repetition and, as some biologists think, a lot of nonsense.

many steps must contribute to building up the genome's large information store. The sum of all their contributions has to account for the whole information gain.

According to the NDT, information can be added only through selection. Selection tests if the mutation is positive or negative, preserving it if positive and destroying it if negative. Even the most complicated mutation serves only as grist for the mill of selection.

Let's look at how an ordinary message, say a telegram, carries information. If the telegram is to carry information, the sender first has to put the information into it. If the sender puts no information into the telegram and sends, say, just random symbols then the receiver gets no information. Also, if the telegram is to bring information to the receiver it's contents must surprise him in some way. If he knew beforehand exactly what the telegram would say, he gets no information.

According to the NDT, the receiver of the information is the genome — not the genome of any one individual, but the average genome of the population. That's where the message is ultimately supposed to be received, and that's where the information is supposed to build up.

When a mutation occurs, selection can choose only between the mutant and the rest of the population. It can choose the better from the good, the more adaptive from the less adaptive. In one step, selection can add no more than one bit of information.[7] That's because it makes only a binary choice between YES and NO, no matter how complex the two options.

Could selection add more than one bit to a genome in one evolutionary step? It can't if the selection is between only two options. If it is to add more than one bit in a step, there would have to be more than two strong options.[8]

How much information is added doesn't depend on how complex the options are. Suppose, for example, a Shakespearean theater group wants to choose a play to put on, and the choice is between *A Midsummer Night's Dream* and *As You Like It*. Suppose also that the group leaves the final choice up to the director who cables them his decision.

His message to them will have only one bit of information no matter how he chooses to express it. His cable could read "THE FIRST," or "A MIDSUMMER NIGHT'S DREAM." In either case his cable message would have the same meaning and carry only one bit of information. Even if he cables them the entire text of *A Midsummer Night's Dream*, his message will have just one bit of information.

If he and they had agreed on a code he could cable his decision with just one binary digit. They could have agreed, for example, that a 0 would stand for *A Midsummer Night's Dream*, and a 1 would stand for *As You Like It*. He could let them know his choice by cabling either a 0 or a 1.

———

Suppose you did an evolution experiment in the laboratory. What would you think if you saw a single mutation lead to a lot of new complexity and new information in the phenotype? As I have already shown you, one step of evolution cannot, on the average, bring to the genome more than one bit of information. You would therefore have to suspect that the complexity was already there. You would have to suspect that the mutation only switched on some complexity that was latent in the genome. The right mutation in a regulatory gene could switch on a whole array of latent genes. A single mutation could in this way bring out a whole complex of activity.

A mutation that reverses the effect of a previous one (called a *back mutation*) could at one stroke revive a complex function that had been earlier shut off. If you didn't know it was a back mutation you might be tempted to think it added a lot of information to the genome. But once you know that a single mutation cannot add more than one bit of information, you know that the complexity must have already been in the genome. The mutation must have turned ON what had been an existing, but dormant, system.

If a copying error were to damage a gene so it no longer functioned, the genome will have become less complex, and some of its information will have been lost. You might think the mutation has wiped out all the information in that gene. You might think that after the mutation, the genetic information is as if the gene weren't there. But the damaged gene *is* still there and nearly intact. The only defect is the one mutated nucleotide.

You can compare the damage in the gene with the damage caused to a mechanical watch if you were to remove a single gear. Without the gear the watch won't work. To an observer who sees only the outside of the watch, the watch acts as if it had no mechanism at all. But the information is not really gone. If the gear were to be replaced, the watch would work again. Similarly, if a mutation damaged a gene, then a back mutation could fix the damage and the gene would work again. So, even if a copying error were to stop a gene from working, the gene would lose only a little information.

The NDT claims to explain how the complexity of life developed. To support that claim it must explain how the information in the genome got built up. You've seen that the NDT requires that each small evolutionary step add a little information to the genome. You've also seen that a single step can't add more than one bit. A step that turns ON a dormant function is not what's needed for evolution as the NDT describes it.* A step that adds no information, or loses information, is also not what's needed. Steps such as these might be included in a chain of Darwinian steps, but they cannot be

---

* A mutation that switches ON a dormant function doesn't demonstrate neo-Darwinian evolution. That's not the kind of evolution Darwin proposed. From where would the neo-Darwinians suggest the dormant genes came? How did they evolve? You could get evolution this way, but it's not the neo-Darwinian kind. I discuss this in Chapter 7.

the typical ones. The typical Darwinian step must add information to the genome,[9] but no more than one bit.

Don't confuse a gain of information in the genome with what's good for the organism. In Chapter 5 I'll show you examples of mutations that lose information and yet benefit the organism in special circumstances. Most of these mutations lose information because they disable a repressor gene.

———

Darwin is said to have made a second important contribution to the theory of evolution. He gave what seemed to be a lot of evidence to support his theory. Since Darwin's time, scientists have brought still more evidence, again with the claim that it supports the theory.

A small change leading to microevolution can appear suddenly with the birth of a single individual. The change can then spread through the population in a few thousand generations.* But the bradys say the large changes of *macro*evolution build up slowly, little by little, as many small changes add up. Large changes are what most people mean when they talk of evolution.

There are good reasons, based on both theory and observation, to believe that microevolution can occur. The change of a single nucleotide can lead to a change in the phenotype that in the right circumstance is adaptive. An example is a mutation in bacteria that results in resistance to streptomycin.

I think Darwin's main contribution to the theory of evolution was the leap he made from *micro*evolution to *macro*evolution. He was a brady, and suggested that a large change is no more than the buildup of many small changes. When a positive mutation appears, according to Darwin, it is selected

———

* If the selective value (SV) is small, as it usually is in the wild, the spread could take a few thousand generations. (Of course, as Fisher [1958] has found, a mutation will not be likely to spread at all unless there were many mutations to begin with.) In laboratory experiments with bacteria, the SV can be made very high. Then the mutation can spread in just a few generations.

and it gets fixed in the population. When enough of these small steps have occurred, the population will have changed a lot from its original form.

Darwin's leap was only a guess, for which he had no evidence. Because macroevolutionary changes need long time spans he could not have had evidence for it. Even today we have none. Was Darwin, and the neo-Darwinians with him, right in suggesting that a large change is built up from many small changes?

The neo-Darwinians understood well why large changes have to be built from small ones. There is a better chance of getting a long series of small changes than of getting one big one in one step. But the relative chance is not the issue. The hard question that tests the validity of the theory is this: Is the chance of building up small changes large enough to make the theory work?

Common sense says that the amazing complexity of life cannot arise out of a random process. The neo-Darwinians use clever arguments to show why evolution should work and why common sense is wrong. One after the other of them has explained that although the variability occurs randomly, the selection process gives it direction and makes it nonrandom. In explaining the role of selection, however, they ignore the main point: Can random changes give natural selection enough of the right genes for evolution to work?

Their arguments are of three kinds: (1) verbal, (2) mathematical, and (3) experimental evidence. That's a mighty array of power to bring for a theory, and if the arguments were solid and correct they should have put the theory on a stable and reliable foundation. The neo-Darwinians would like everyone to believe they have done that.

(1) Verbal arguments should always be suspect. Clever debaters have long shown they can make even the weakest case look strong. Many trial attorneys make their living doing just that. Darwin used verbal arguments with great success, even though the arguments were specious. As W. R. Thompson, a

highly regarded biologist and a Fellow of the Royal Society, has said,

> Darwin did not show in *The Origin* that species had origi-
> nated by natural selection; he merely showed, on the basis
> of certain facts and assumptions, how this might have hap-
> pened, and as he had convinced himself he was able to
> convince others. [Thompson 1963, p. xii].

Gertrude Himmelfarb has shown how Darwin was a master of the technique of verbal argument in urging the acceptance of his theory. She pointed out how Darwin exploited to his advantage the admitted difficulties of the theory. He con-tended that his theory explained them and, therefore, they could be used as arguments *for* the theory.

> This procedure, by which one of the major difficulties of
> the theory was made to bear witness in its favor, can only
> be accounted for by a confusion in the meaning of
> "explain" between the sense in which facts are "explained"
> by a theory and the sense in which difficulties may be "ex-
> plained away." It is the difference between compliant facts
> which lend themselves to the theory, and refractory ones
> which do not and can only be brought into submission by a
> more or less plausible excuse. By confounding the two,
> both orders of explanation, both orders of fact, were en-
> tered on the same side of the ledger, the credit side. Thus
> the "difficulties" he had so candidly confessed to were con-
> verted into assets. [Himmelfarb 1962, p. 334][10]

(2) The mathematics laid down by Ronald Fisher and Se-wall Wright were meant to show that Mendel's results did not contradict the NDT. Fisher and Wright were also con-cerned with the practical problems of cross breeding crops and cattle. But they did not address what I present here as the main current theoretical problem of Darwinian evolution. They did not look into how likely are the events the NDT says are random.

A theory built on random events must be checked against the probabilities of those events — that's the first check that should be made. Why wasn't that check made? When the theory was being established in the 30's and 40's, the molecular basis of mutations was unknown. The DNA as the repository of genetic information was not discovered until a decade later. Indeed, the neo-Darwinians did not think there was a problem with the probability of mutations. But there is a big problem with it.

(3) The experimental evidence should be the most telling of all in favor of a theory. Unfortunately, there is no direct experimental evidence of large-scale evolution. Of course, there is fossil evidence, but at best the fossils only show that there have been changes in living organisms in the past. They don't tell us how those changes took place. They don't even tell us that the later forms of life descended from the earlier forms. To say that they did descend is an inference that must depend on a theory. So we're back to the question of whether the theory is any good. One cannot then say that the fossils support the theory unless we beg the question and assume the theory to be correct.

These are the arguments that are given to show why we should forego what our common sense tells us and believe in descent. But, as we shall see, common sense is still a reliable touchstone, even in evolution.

NOTES TO CHAPTER 3

1. **Survival of a mutant**. How do random effects tend to wipe out a mutant even if it has a selective advantage? If a population is stable its numbers don't change. That means each member will, on the average, have exactly one surviving offspring that will replace it, or each mating pair will have two surviving offspring to replace them. Any particular individual may, of course, have many offspring or none at all, but on the average there will be exactly one surviving offspring per individual.

A population can stay the same size only if there is an average of one reproducing offspring per member of the population. The offspring that don't reproduce don't count. Some members cannot reproduce. Others that can reproduce might not do so because they die early. They may die for any of several reasons. Some may be eaten by predators and some may die in a catastrophe such as a flood or a fire. Often, they die not through any fault in their own abilities but through random acts of fate.

To ensure that there will be a survivor to reproduce, an organism gives birth to more than one offspring. An elephant in the wild, may have from five to ten offspring during its lifetime. Frogs can lay hundreds of eggs at once, and they will repeat this several times during their lifetime. A herring will lay about 50,000 eggs per year. An oyster can produce 100 million eggs in its lifetime. A lower plant can produce a trillion spores. If the population is stable, an average of only one of these calves, eggs, or spores will live to maturity and reproduce.

Suppose an animal in a stable population produces, on the average, five offspring in its lifetime. Since the population is stable, an average of only one out of five will survive to reproduce. That means the chances are only one out of five, or 20 percent, that a particular animal at birth will survive to reproduce. Whether or not it survives is largely a matter of luck. Its own abilities often have little to do with it. A mutant with a selective value of 0.1 percent will have only a slightly higher chance of survival. Its chance to survive is 0.1 percent higher than the others, which means that its chance to survive is 20.02 percent instead of 20 percent.

A single mutant with an *SV* of 0.1 percent may or may not survive. The chance would be 0.2002 that it would survive in the first generation. The chance would be 0.1347 that it would survive 2 generations. The chance that it would survive the third generation is still less. Its chances of survival in the first few generations are not affected much by its selective value. The chance of any of the others in the population having a descendant that survives the first generation is 0.2000. Their chance of surviving the first two generations is 0.1345. Notice that I have to use four significant figures in the probability of survival to see a difference between the mutant and the others.

The chance that one positive mutant will survive is not much different from the chance of the nonmutants surviving (it only differs by a tenth of a percent). It, too, can suffer a random death even though it has a positive selective value. If its selective value were ten times higher, namely 1 percent, its chance of surviving the first generation would be only 0.202. So if a single mutant appeared, its descendants could vanish through random effects much like the descendants of any of the others. Its positive selective value would not help it much to survive the random effects that tend to destroy it.

So in our example, the odds are about 5 to 1 against a single mutant surviving to have offspring. It has to be lucky to survive and have its descendants take over the population. For an adaptive mutant to have a good chance of survival there have to be many of them in the population even though their fraction may be small.

The survival of a mutant in a population is like a game of chance. The selective value of the mutant does not ensure it will survive and take over the population. The selective value only affects the *chance* it will do so. The larger the selective value the better the chance it will survive. A mutation with a large selective value may occur, but that's not the end of the matter. It still may not get selected. It's all a matter of chance.

Fisher's method takes these effects into account in finding the chance that a mutant will survive.

2. **Survival of any one of several mutants**. If the chance of survival of one mutant is 0.002, you might think the chance of survival of either of 2 such mutants is twice that, or 0.004. If you thought so you would be close to correct, but 0.004 is not exact. You might then go on to think the chance

of survival of any one of 10 such mutants is ten times 0.002, or 0.02. If you thought this you would again be close to correct, but not so close as you were before. If you tried to find the chance of survival of any one of 1000 such mutants by multiplying the number by the chance of survival of one of them you would see clearly why the method is incorrect. Multiplying 0.002 by 1000 gives a chance of 2, and that is surely wrong. A chance of 1 is certainty. There is no such thing as a chance of 2.

The proper way to find the chance of survival of at least one out of 2 is to note that since the chance of the first one surviving is 0.002, the chance it will disappear is one minus 0.002, or 0.998. The chance the second will disappear is also 0.998. So the chance they will both disappear is 0.998 × 0.998 or 0.996004. Then the chance that at least one will survive is the chance they will not both disappear. That's 1 - 0.996004, or 0.003996. That's close to the value 0.004 that you would have guessed. The chance that at least one of 500 such mutations will survive is $1 - 0.998^{500}$, or about 0.632 (about 5/8). The chance that at least one of 1,000 such mutations will survive is $1 - 0.998^{1,000}$, or about 0.865 (about 6/7). The chance of at least one out of 2,500 surviving is $1 - 0.998^{2,500}$, or about 0.993.

3. **The size of the population affects the chance of survival.** When Fisher solved for the chance of survival of a mutant he took the population to be infinite. I solved the problem of gene survival in a finite population in an approximate way by taking it to be a random walk with two absorbing barriers [Feller 1957]. I then found the chance of survival to be $2S/(1-e^{-2SN})$, where S is the selective value and N is the population size. Note that the larger N is, the smaller is this chance. The chance is always bigger than $2S$, but when SN is large, the chance is close to $2S$. When SN is 1 the chance is about 16% larger than $2S$. When SN is 2 it's about 2% larger. When SN is 3 it's only about 0.25% larger.

4. **Dividing a population**. If you divide a population into smaller groups you'll raise the chance that a positive mutant, if it occurs, will survive. The mutant will survive better in a small group than in a large one. That follows from the expression I gave in the previous note for the chance of survival.

But as you can see from the numerical values in the previous note, reducing the population gains only a little in the chance of survival. The population has to be small to get a significant increase in the chance of survival. To get substantial benefit from this effect you have to make the population very small, but then the whole group is in danger of extinction by a minor catastrophe.

5. **Limits on mutation rates.** The mutation rate an organism can tolerate depends on how many essential nucleotides it has in its active genome. If it has 100 million essential nucleotides, it cannot tolerate a rate of much more than one in a billion per reproduction. At that rate, about a tenth of the population would suffer genetic damage. A much higher mutation rate would ruin the genome of most of the population. It would change some of the essential information in it. Single-celled organisms can tolerate higher mutation rates than plants or animals because they have smaller genomes.

6. **The variability number**. I got the number $10^{24,082,400}$ for the variability as follows: The genome of a mammal has about 4 billion nucleotides. I took a conservative view of the information content of the genome. I assumed, as do many molecular biologists, that only about 1% of the DNA is real information, and that the remaining 99% carries no information. (This may not be true, but if more than 1% of the genome carries information, the variability number will be even larger than the number I am using.) I assumed then that all the information is in 40 million nucleotides.

There are four different DNA bases. There are therefore four ways of choosing each nucleotide of the forty million. There are four ways of choosing the first nucleotide. For each of those ways there are four ways of choosing the second. There are then 4×4, or 16, ways of choosing the first two. So there are 16 possible sequences of the first two nucleotides. There are 4×4×4, or 64, possible sequences of the first three nucleotides, and so on. The number of ways to choose a string of forty million nucleotides is 4 raised to the power 40 million, which is about $10^{24,082,400}$.

7. **Information in a binary choice**. The information in a binary choice is at most one bit. You get the full bit only when the chances of getting the two options, call them ZERO and ONE, are the same. If there's a bias toward either option, the information in the choice will be less than a full bit. You can see that if you look at the extreme case in which the bias is so strong that ZERO is certain to be chosen and ONE is certain not to be. Since there is no uncertainty the recipient of the message could predict beforehand that it was going to be a ZERO, and the message would give him no information.

In a case short of the extreme, where the ZERO is not certain but only has a better chance to be picked than the ONE, the information is larger than zero but less than one bit. In any case, a binary choice cannot be more than one bit.

8. **No more than one bit of information per step**. You might want to say that in one step there could be more than one positive mutation in the population. You might think that, if there were more than two alternatives, more than one bit of information might be generated by selection. That might be so if selection were strong enough always to pick the mutant with the largest *SV*. But you have seen that selection is chancy. Unless a mutant has a large selective value, the chance of it surviving is small. So whatever mutants there are with a positive *SV* will together play the role of one alternative, and the rest of the population will be the other. When selection chooses between these two it generates no more than one bit of information.

9. **How can information be added to the genome?** Adding information could mean adding new DNA symbols,

but that's not the only way to do it. Indeed, you could add information without adding symbols at all. On the other hand, you could add symbols to a message without adding information. How?

You can easily add symbols to a message and not add information: just add random symbols. Then you won't be adding information — you'll be adding only nonsense. Similarly, if you add random nucleotides to the genome you add no information. Symbols without meaning carry no information.

It's a little harder to add information without adding symbols. To do it you have to start, for example, with a message that already has some redundancy, such as a string of nonsense symbols. You can then add information if you replace the nonsense symbols with some that carry meaning. If you want to add information without adding symbols you have to start with some redundancy. You have to start with a message whose symbols don't carry information at full capacity. You have to start either with symbols that aren't carrying information at all or that aren't carrying as much as they could.

The neo-Darwinians say that information could have got into the genome in the following way. Somehow, extra nucleotides got into the genome. They could have got there by duplicating existing DNA or by random DNA entering somehow. If this added DNA is not functional, it's free to change without causing any harm. Random mutations in this free DNA might by chance convert it into a working gene and have a favorable effect on the phenotype. Natural selection, acting on the phenotypes, will favor some mutations and reject others. The phenotypes will, of course, carry along their genotypes. As the population evolves, its genome gets more and more complex. More and more information builds up in it a little at a time. In this way, they say, the large amount of information and complexity in living organisms has been built up.

10. The nature of Darwin's arguments. Darwin recognized that the fossil record did not support his theory (as it still does not). He examined several cases of specialized forms of animals where he admitted the difficulty of explaining how such a form might have arisen. He took up several such refractory examples and dealt with them all in much the same way. In the case of the bats, for example, he admitted that the problem of bat origin is a hard one. Then to solve the origin of the bats he started with

> the Galeopithecus or so-called flying lemur, which formerly was ranked amongst bats, but is now believed to belong to the Insectivora. ... Although no graduated links of structure, fitted for gliding through the air, now connect the Galeopithecus with the other Insectivora, yet there is no difficulty in supposing that such links formerly existed ..."
> [Darwin 1872, pp. 182 ff.]

From "there is no difficulty in supposing", he went on to lead the reader into agreeing that it *probably* happened, and finally he assumes the reader goes along with him in agreeing that it did indeed happen that way. His arguments, however, consist of no more than " I can see no difficulty ...", to "there is no difficulty in supposing ...", "nor can I see any insuperable difficulty in further believing ..." which he parlays up to "it is conceivable that ..." and to "we might expect that ...". What started out as a difficulty became, at the end, another "demonstration" of the power of natural selection.

Interestingly enough, even today, the origin of the bats is not found in the fossil record. Robert Carroll, a paleontologist from McGill University and curator of vertebrate paleontology at the Redpath Museum has noted that there is no evidence of the origin of the bats in the fossil record. The earliest fossils of bat skeletons found are "almost indistinguishable from living bats." [Carroll 1988]. Moreover, the sophisticated sonar-like echolocation system of modern bats has been found as well in the oldest known bats [Novacek 1985].

Chapter 4

IS THE DECK STACKED?
Can the variation be random?

THEY called him *Chuck*. That wasn't his real name. His real name was *Francis*, but woe to anyone who would call him that. He was tall, his face was weather-beaten, his hair was black with just a touch of gray at the temples. You couldn't see much of his hair, though, because he always wore his Stetson low on his forehead shading his eyes. The hat together with his tan shirt, the blue bandanna around his neck, and his brown leather shop-made boots made him look very much the cowboy that he was.

He was taciturn. He never spoke unless he had to, and he never made idle conversation. He had a reputation for being fair and honest. He was even tempered; he wouldn't anger easily. Even on those rare occasions when he did get angry, his face didn't show any emotion. He would vent his anger in a cool and calculated way.

Chuck rode in to Abilene the day before on the Chisholm trail, and he was flush with money. He had driven the herd all the way up from Texas, and now he wanted to relax. He relaxed best by playing poker. He was a natural-born poker player; no emotion ever showed.

They were playing straight poker. Chuck was winning a little, and he was feeling pretty good. Bert, the New Yorker sitting across from him, was the only one at the table Chuck didn't know. Bert was the opposite of Chuck. He was a salesman, and a source of constant babble that Chuck hated.

Bert was taking his turn to deal the hand. Chuck picked up his cards slowly, one by one. He saw a ten of diamonds, then a seven of diamonds, and a jack of diamonds. It could be a possible flush, maybe even a straight flush. Then came a nine of diamonds! Could it be? He was almost afraid to pick up the last card. But he kept his cool and didn't break his pace. His right hand continued its smooth motion as he placed the last card with the others in his left hand. It was a fantastic hand of cards! IIe had a straight flush to thc jack! Hc showcd no emotion. But he felt elated. He felt lucky. When the betting got to him it was at a modest five dollars. He raised only another five. He didn't want to discourage the others too early.

Bert raised ten dollars. Sam stayed and so did Ed. Chuck raised only another five dollars. There was no point in scaring everyone out. He had plenty of time. The betting was unlimited — not exactly a gentlemen's game.

Fred, to Chuck's right, dropped. Don called. Bert raised twenty dollars. Sam called. Ed dropped. Chuck raised another twenty. Both Fred and Don dropped. Bert raised fifty. Sam, perspiring, hesitated for a long half minute, then he threw his cards face down and dropped. Chuck raised a hundred; now he was going for it. Only he and Bert were in.

Bert raised five hundred. Chuck was surprised. He raised another five hundred. That was all the money he had, but this hand was money in the bank.

Bert raised a thousand. Chuck raised another thousand (which he didn't have, but his credit was good). What could Bert have in his hand?

Bert laughed and raised ten thousand. Well, that was steeper than Chuck ever would have gone. Bert must be trying to force him out. Chuck knew he had to see him. Chuck went into the hole for the ten thousand and called for the showdown, wondering what Bert could have.

With a laugh Bert put down his cards slowly one by one: nine of clubs, jack of clubs, king of clubs. Looks like Bert had a flush. Bert went on and put down a queen and a ten of clubs! He had a straight flush too, and to the king! He was still laughing when Chuck pulled the trigger and sent a single thirty-eight-caliber bullet through his forehead.

"Chuck!", cried Ed. "What the hell 'djya do?"

"He was cheatin'," Chuck said softly, as he slowly returned his gun to its holster.

"How d'ya know that?" Ed shouted. "That's not like you! Ya' don't kill a man in cold blood?"

He repeated, barely audible, "He was cheatin'."

"If ya' saw 'im cheatin', why 'dja lett'im get that far? Why didn'tcha stopp'im before?"

"Ah didn't know he was cheatin'," he said slowly, looking away, and barely above a whisper, "until he showed his straight flush." He looked Ed straight in the eye. "That ain't jest luck."

"How d'ya know it wasn't luck! You can't jest kill a man like that!"

"He set me up," Chuck said in a slow deliberate tone without emotion. He looked up at the ceiling and paused. "Ah reckon the odds against both of us gettin' a straight flush are more'n a billion t'one. If sump'm happens against odds like that," he turned, looked at Ed, and spaced his words, "it .. ain't .. luck!"

"Yer crazy! Yer positively crazy! Ya' can't kill a man like that!"

"Ah can't be wrong very often," he said turning to stare out the window. "Not more'n once in a billion, anyway." After a pause he added thoughtfully, "An honest man'll sooner get struck by lightnin' than get shot by me."

In this chapter we are going to see if the genetic variations needed by the neo-Darwinian theory (NDT) can be random. Could they have come by chance? Are the adaptive mutations that are supposed to make up macroevolution just luck?

Were the poker hands just luck? Or did Bert stack the deck? There are two ways we can approach a problem like that. We can try to gather evidence that Bert cheated. Maybe someone saw him cheat. Perhaps we can build a case from what people know about Bert's past behavior. Or we can compute the chance of getting two straight flushes in one hand, like Chuck did. Then we can try to come to a conclusion that is unlikely to be wrong.

Let's approach the problem of the randomness of the mutations in much the same way. If we should find overwhelming evidence that they are not random, then we shall have shot down one of the two important pillars of the NDT.[1]

The tachys say that a macroevolutionary change is more often a single large random change than it is a chain of small ones. They say that large changes in the phenotype come mainly from mutations in regulatory genes.

John McDonald, of the University of Georgia, is a tachy. He favors insertions of DNA segments, known as RLE's (*retroviral-like element*), into regulatory genes as the mutations that drive evolution [McDonald 1990]. RLE's are transposable elements, or transposons, that are like a special kind of virus, known as a *retrovirus*.* A retrovirus has a genome of double-stranded RNA. When it gets into a cell, it copies its RNA onto a single strand of DNA with an enzyme that it carries, called *reverse transcriptase*. Then it uses the same enzyme to copy the DNA to get a second strand of DNA. Using another enzyme (which, as far as I know, hasn't yet been isolated) it inserts this double-strand of DNA into the cell's genome. Each step is controlled by special enzymes.

An RLE contains special pieces of DNA on each end known as LTR's (*long terminal repeats*). The LTR's themselves contain genes that regulate the rest of the RLE. The main part of the RLE contains genes for transposing the RLE

---

* AIDS is caused by a retrovirus.

as well as genes that can regulate other parts of the genome not in the RLE [Stryer 1988].

But what kind of regulatory changes do the tachys suggest? All they have suggested so far are ways of turning existing genes OFF and ON. The insertion of a DNA segment into a gene can indeed turn that gene OFF and keep it OFF in future generations. An inversion in a gene will do the same. But how will the gene be turned ON? To turn it ON, another insertion or inversion must occur that precisely undoes the first one.

Two questions arise here. The first is, can these insertions and inversions be random? The second is, how much information can mutations like these add to the genome?

There are good reasons to believe that these genetic rearrangements are not random. Insertions and inversions are complex rearrangements of the gene. Inversions occur when two sequences recombine in just the right way. Inversions seem to have important roles to play in both cells and organisms, but we don't yet know what those roles are. We do know, however, that they are not just genetic mistakes. They are controlled by a set of special enzymes [Darnell et al. 1986]. Some of the enzymes they need are encoded in the transposon itself, and some are encoded in other parts of the cell's genome. The rearrangements seem to be deliberate acts performed on behalf of the cell (or the organism). They do not seem to be the random stuff that the NDT says propels evolution.

Insertions and inversions can disable the gene into which they enter by disrupting the reading frame. In this way, they act as switches to turn a gene OFF. The insertion can also be precisely removed, and the inversion precisely reversed, to permit the gene once more to function. The insertion sequences (IS) have special indicators on their ends that identify them for removal and transposition. If it were not for the precision with which they act, they would be turning genes OFF at random, wreaking havoc in the genome. Moreover, if not for the special indicators on the ends of the IS, once a gene is turned OFF in this manner there would be little chance the IS could be precisely removed to turn the gene ON again. The chance that a random deletion will precisely take out a previous insertion is very small.

The chance is also small for a random inversion to reverse a previous inversion. The chance that a transposition will occur in the genome of a bacterium is about one in a million generations [Stryer 1988]. If they are random, then they would be equally likely to occur anywhere in the genome. Recall that the genome of a bacterium has about a million nucleotides. Then the chance that an inversion or a deletion will occur and have one of its endpoints on a particular nucleotide would be about one in a trillion replications. Although we might expect to see such events in bacteria, we couldn't expect them to be important in higher animals because mutations in them are not frequent enough. During the whole of the estimated 65 million years of horse evolution, for example, there have been only about a trillion replications [Simpson 1953].

Moreover, the tachys' suggestion does not account for how information can build up in the genome. If they turn ON a regulatory gene, they can bring into play a complex function, or even a whole system of functions. But, as I noted in Chapter 3, the information must already be in the genome. It is just waiting to be turned ON. With this we see that genetic rearrangements cannot serve as the random variation required by the NDT.

Let me say here that I am inclined to be more of a tachy than a brady. I think the mutations that drive evolution are mainly in regulatory genes. They might include the genetic rearrangements described above, but we don't know enough about them yet. I think these mutations lead to large changes in the phenotype, but they cannot be random. To compare them with the cowboy's poker game, I would say they're not just the result of luck. They're dealt from a stacked deck. My thinking on this point is very different from the NDT, and therefore from the Darwinian paradigm. I'll return to this point in Chapter 7.

———

The bradys hold that a large evolutionary change occurs through a long chain of small steps. This is the process we have called cumulative selection. They hold that the mutations in these small steps are the copying errors. Everyone agrees that the copying errors are indeed random. With the discovery of copying errors, there was a wide consensus that they were the ultimate source of the required variation. They don't need any special cell machinery, and they are random as the theory requires. So what's the problem with cumulative selection? Why can't it be the mechanism that drives evolution?

I have already pointed out some of the things wrong with the bradys' concept of macroevolution. These things were, in fact, why some evolutionists became tachys — and those reasons are valid. The chance of getting the necessary mutations is just too small if it's done through cumulative selection. Much less probable than two straight flushes in one poker hand, they're just too lucky for anyone to believe they really came by chance. To understand why this is, we shall have to do a little arithmetic, because the degree of luck depends on the numbers.

For cumulative selection to work, a lot of good mutations have to occur by chance. At each step of cumulative selection, a mutant with a positive selective value has to appear. It also has to be lucky enough to survive and eventually to take over the population. Then another good mutation has to appear for the next step, and so on. The neo-Darwinians seem to think the chance of all this happening is large enough to make evolution work. But no one has ever shown that to be so. No one has ever shown that such a thing is likely — or even possible!

Until the latter half of the twentieth century, this kind of check couldn't be made. Only since the middle of the twentieth century have we begun to understand what the genome is. Only since the 60's have we been able to estimate the chance of a mutation. I made some of those calculations [Spetner 1964, 1966, 1968, 1970].

A change of a DNA nucleotide affects the protein or the RNA encoded by the gene. If it's a protein, it might be an enzyme, or it might be part of the structure of the organism. It might help control the making of other proteins, or it might help control development. A change in a protein can lead to a change in the phenotype.

The rarity of copying errors is a problem for the NDT. The average rate of copying errors depends on the kind of organism. In bacteria the mutation rate per nucleotide is between 0.1 and 10 per billion transcriptions [Fersht 1981, Drake 1969, 1991]. But in all other forms of life the rate is smaller. For organisms other than bacteria, the mutation rate is between 0.01 and 1 per billion [Grosse et al. 1984]. The geometric-mean[*] rate is one per billion (10^{-9}) in bacteria and one per ten billion (10^{-10}) in other organisms. These are the chances of a mutation in a particular nucleotide in a particular replication. I'll use these mean rates in my calculations.[2]

Error rates are as low as they are only because the cell has a proofreading mechanism which corrects most of the errors made in transcription. The evolutionists say this mechanism evolved as everything else did. It wouldn't have developed if it weren't necessary. Proofreading keeps the rate of mutations low enough so they won't be likely to spoil a genome that's already working well. Lubert Stryer of Stanford University has suggested that the mutation rate has been optimized by evolution. He said:

> ... the mutation rate of a species has been optimized in the course of evolution. Too high a rate would lead to nonvi-

---

[*] The geometric mean is the mean on a logarithmic scale.

able progeny, whereas too low a rate would diminish genetic diversity [Stryer 1988, p. 676].

So when I consider animals, I'll use the mean (geometric) mutation rate of one in ten billion (10^{-10}). These rates are about as high as they can be.

When I compute the values of the chances (or probabilities) of some of the events of evolution, I'm going to get some very small numbers. They're so small that we don't have an intuitive feel for their size. To gain a feel for these small chances let's first look at some familiar events of low probability.

To win first prize in the New-York-State Lottery you have to choose correctly a set of six numbers from 1 to 54. The chance of winning is about one in 26 million. That's not the chance that *some* set of numbers will win. It's the chance that *your* set will win.* The chance that *someone* will win depends on how many tickets are sold, and that chance is fairly high.

The chance that in a bridge game you will be dealt 13 cards of one suit is one in about 160 billion (1.6×10^{11}). The chance that all four players at your bridge table will each be dealt 13 cards of one suit is one in about 2.2×10^{27}. That's about a hundred thousand times less likely than your set of six numbers winning the New-York Lottery three times in a row. It's about ten times less likely than having your number come up on a roulette wheel 17 times in a row.

But even if you should win, you'd have a hard time collecting first prize in the New York lottery three times in a row. Nor would you be allowed to win at the roulette table 17 times in a row. Such events are so unlikely that everyone would suspect fraud. Regardless of any precautions that might have been taken, fraud is far more likely than that kind of luck. Nearly everyone would think such luck to be so im-

---

* The purchase of one ticket for a dollar gives you two chances; you choose two sets of six numbers. The chance that *either one* of your two sets of numbers will win is about one in 13 million.

possible that there would have to be something wrong. As we've seen in the story of the poker-playing cowboy, too much luck might not be good for you. So too in nature, if we see the occurrence of an event with exceedingly low probability, we must suspect the event was not random and that we were mistaken about its low probability.

We often use the word *impossible* to mean a very small chance, or something *very improbable.* Just how improbable an event must be to be called *impossible* depends on who's judging. Cowboy Chuck felt anything less probable than one in four billion was impossible. He was even ready to pull the trigger of his six shooter to back that feeling. I would suspect the New-York-Lottery authorities would feel that winning two weeks in a row, which is a chance of about one in a hundred trillion, is impossible. I doubt if they would pay off the second win.

Let's look at an event that almost everybody would call impossible. Consider the very unlikely event of flipping 150 coins and having them all come up heads. This event will have a chance of one in 2^{150}, or one in about 10^{45}. So you would have to flip the 150 coins about 10^{45} times before you could expect to see all heads. To gain some feel for the chance of this event let's see how many times you could flip 150 coins.

What kind of an operation would you have to mount to flip 150 coins 10^{45} times? Flipping real coins is slow and cumbersome. You not only have to flip them, you also have to count the heads and pick them all up for the next flip. You couldn't do it even if you devoted your life to it. Suppose you were super fast and could flip 150 coins, count the heads, and pick them up, all in one second. There are only about 3 billion seconds in 100 years. Even if you had a thousand super fast people help you, each flipping the coins once a second for a hundred years, you could only flip three trillion times. That's only 3×10^{12}, and it's a long way from 10^{45}.

So let's think of doing it faster by simulating the flips on a computer. Suppose you could have a technology more advanced than today's, and that you could simulate a flip of 150 coins in a trillionth of a second. (That's the time it takes light to go about a hundredth of an inch.) Then build a computer made up of a billion of these simulators.

Even at this speed and even with such a big computer you couldn't do it alone. You'd need a lot of help. So make 10 billion of these computers and assign one (or two) to each person on earth to monitor. Let the experiment run for 3,000 years (beyond that the human race might lose interest in the project). In all that time and with all those computers you could make only 10^{42} simulated trials.

In 10^{42} trials, the chance that at least one will be all heads is $10^{42}/10^{45}$, or 10^{-3}, or one chance in a thousand. There's one chance in a thousand that, with 10 billion big computers each simulating a billion flips of 150 coins every trillionth of a second for 3,000 years, at least one of the flips will be all heads. You would never expect to get all heads in only one flip of those coins. With good reason, then, you could say that in a practical sense it's impossible to get all heads when you flip 150 coins once.

You wouldn't expect to see such an event in your lifetime. Nor would you expect anyone on earth to see it in his lifetime. Nor would you expect any human being who has ever lived to see such an event in his lifetime. That should qualify the event as impossible enough for you to say, if you heard of it happening, that it could not be luck. You would not brook a theory that explained such a happening as just luck.

The kind of events that are supposed to run evolution take millions of years and large populations. There can't be many trials of this magnitude from which to find a winner. I'll compare the chances of some of the events of evolution with this event, which most people would call *impossible*.

Now let's estimate the chance of the kind of event that is most important in evolution: the emergence of a new species.

For now, let's look at evolution through the eyes of a brady. What is the chance of the whole series of steps occurring?

To calculate, we have to know:

- What the chance is of getting a mutation

- What fraction of the mutations have a selective advantage.

- How many replications there are in each step of the chain of cumulative selection.

- How many of those steps there have to be to achieve a new species.

If we get values for these parameters we can find the chance of evolving a new species.

We already have the first of these parameters the mutation rate. The mean mutation rate for animals is 10^{-10}.

Note that not any copying error can serve a typical step in cumulative selection. To be a part of a typical step a mutation must:

1. have a positive selective value, and

2. add a little information to the genome.

Although both these requirements are obvious, I don't know that anyone has ever pointed out the second one before.

How small can a mutation be and still qualify as a typical step in cumulative selection? No one knows. But because of the above two requirements it can't be as Richard Dawkins [1986] has assumed — it can't be as small as you please and still qualify.

Can these mutations each be a change of just one amino acid in a protein? Can all the mutations be as small as that and still have selective value and still add information to the genome? If not, can they be a change in two amino acids? In three? In ten? No one knows.

The minimum change in a genome is a change in one nucleotide.* Let's try to find the chance of this minimum mutation and see where it leads. Note that we don't know if mutations can all be of the minimum size of one nucleotide and still satisfy the above two requirements.† But let me assume it is possible so I can proceed, even if it means giving away this point to the evolutionary side of the argument.

How many small selective steps would we need to make a new species? The smaller the change in each step, the more steps we would need. G. Ledyard Stebbins, one of the architects of the NDT, has estimated that to get to a new species would take about 500 steps [Stebbins 1966].‡

How many births would there be in a typical small step of evolution? Paleontologists have been studying the evolution of the horse for a long time, and they think they understand it well. From the fossil record they have inferred changes in the horse during the past 65 million years. Some examples of these changes are that it grew larger, it got stronger legs, harder teeth, and a larger brain. Using the numbers cited by the experts, I find that one small evolutionary step would comprise about 50 million births.[5]

I still need a figure for the fraction of mutations that are adaptive. How many mutations would have a selective advantage? Unfortunately, no one knows the answer to that ques-

---

*A change in one nucleotide of the DNA does not give so much freedom as a change of an amino acid of a protein. A change in a single nucleotide can change an amino acid to one of at most *three* others. Some nucleotide changes can effect a change to only one of *two* other amino acids, some can change to no more than *one* other, and some nucleotide changes won't change an amino acid at all.

† None of the point mutations that have been observed satisfy these requirements.

‡ I am trying to be realistic in using the figure of 500 steps to achieve a new species. Of course, the larger I make this number, the less likely we shall find evolution to be. My justification for the use of 500 steps is Stebbin's estimate, which is meant to represent a typical case of cumulative selection. I am here trying to see if a long chain of point mutations can explain macroevolution.

tion. And without an estimate of this fraction I can't go on with the calculation.

So I shall turn the calculation around. How many mutations must be adaptive for the NDT to work? How many must there be to make the chance large enough of getting a new species?

Let's start by computing the chance that a copying error occurs at a particular place on the genome of one individual and then spreads through the population. Once I find that chance, I can find how many potential adaptive copying errors there must be to make the theory work. Let's see if this number turns out to be reasonable.

What I mean when I say "to make the theory work" is that cumulative selection should lead to a new species by successfully completing 500 steps. But the completion of these steps is a random event — it's a matter of chance. We can only calculate the chance that it will occur. We shall have to adopt some level of chance of achieving a new species as our criterion that the theory works. Then we can ask how many potentially adaptive mutations there must be to get to that level of chance.

We want to set a reasonable level of chance that the NDT work. The higher we set that level, the harder it will be for evolution to work. The lower we set it, the easier. Surely the chance doesn't have to be so large as 1. Asking that the chance be 1 of achieving a species would be to require that it be certain. It's reasonable to ask that the chance be less than 1. Not every species has to evolve for the theory to work. Some species can become extinct. How much less than 1 should we ask that it be?

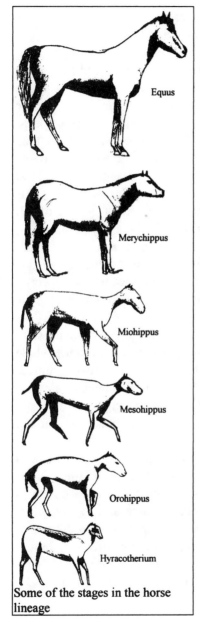

Some of the stages in the horse lineage

Richard Lewontin of Harvard University has estimated that for each species alive today there are about 1000 that went extinct [Lewontin 1978]. Shall we then say that for evolution to work, the chance of getting a new species should be one in a thousand? On the one hand, maybe it should be larger. Some of those extinct species are not without living descendants. Some of them became extinct because their descendants evolved into species that replaced them. On the other hand, some species are around today that do not seem to have evolved for a long time. So maybe the chance of speciation should be smaller than one in a thousand.

Some species go for a long time without changing. This is the stasis that the punctuationists have emphasized. So let's throw in another factor of a thousand for this effect.

Let's then set the level of chance to one in a million. Thus we adopt the criterion that evolution can work if the chance of achieving a new species in 500 steps is at least one in a million. If the chance is less than that, we shall say that evolution does not work.

If the chance of getting to a new species (a full 500 steps) is one in a million, the chance of each step must be much larger. An adaptive mutation has to occur and take over the population at each of the 500 steps with a chance large enough to make the chance of all 500 of them no smaller than one in a million. The chance of a single step has to be so large that when we multiply it by itself 500 times we get at least 1/1,000,000. The smallest number that will do that is close to 0.9727, which is a chance of about 36 out of 37. (You can check that with a pocket calculator if you like.)

Now let's find the chance that a mutation in a particular nucleotide will occur and take over the population in one step. What's the chance that a mutation occurs in a specific nucleotide of the genome during one evolutionary step? The chance of a mutation in a specific nucleotide in one birth is 10^{-10}, and there are 50 million births in an evolutionary step. The chance of getting at least one such mutation in the whole step is about 50,000,000 times 10^{-10}, or one in two hundred. There is an equal chance that the base will change to any one of the other three.[*] Then the chance of getting a specific change in a specific nucleotide is a third of that, or one in six hundred.

Note that I have taken the mutation at each step to be a change in a single nucleotide.[6] I don't know if there is always, at each stage, a single nucleotide that can change to give the organism positive selective value and to add information to it. No one really knows. But I have to assume it can if I am to get on with this study of cumulative selection.

That's a pretty strong assumption to make, and there's no evidence for it. But if the assumption doesn't hold, the NDT

---

[*] The chances aren't really equal, but assuming they are will give us a result that is close enough for our purposes.

surely won't work. Although we don't know if it holds, let's see if the NDT can work even *with* the assumption.

Most popular descriptions of the NDT imply, if they don't outright say, that even the slightest improvement resulting from a mutation would play a role in evolution. I've already cited Darwin and Dobzhansky on that. Dawkins [1986, p. 49], has also made that claim in describing cumulative selection as a process "in which each improvement, *however slight*, is used as a basis for future building" (my emphasis). That statement is, of course, false. Most "slight improvements" won't be a basis for anything. Unless they occur in large numbers they will most likely disappear. As Fisher has shown, it's a matter of chance that a mutant survive. It might spread through the population and take it over, but more likely it will just vanish. In fact, the slighter the improvement the greater the chance it will vanish.

The mutation that leads to the improvement must also be *dominant*. That means that the mutation will be expressed in the phenotype even if it's on only one of the two chromosomes that carries the gene. If it had to be on both chromosomes to be expressed, which is the condition we call *recessive*, then the mutation would have to occur in both a male and a female and the two would have to find each other and mate. This last condition is much less likely than the mutation appearing just once.

Since the mutations I must consider are small (indeed, they are the smallest possible) any selective value they have must also be small. Fisher's analysis shows that a mutant with a selective value of one percent has a two percent chance of survival in a large population. That's a chance of one in 50. If the selective value were a tenth of a percent, the chance of survival would be about 0.2%, or one in 500. If the selective value were a hundredth of a percent (0.0001) the chance of survival would be about 0.02%, or one in 5,000.* For large

---

* This figure holds for a population of 10,000 or more. For a population of 5,000 the chance would be closer to 0.0003.

populations, the chance of survival turns out to be about twice the selective value.

Sir Ronald Fisher did much of the original work in population genetics, and his work is still the standard in the field. He found from his studies that even good mutations are likely to disappear from the population. He said:

> A mutation, even if favorable, will have only a very small chance of establishing itself in the species if it occurs once only. [Fisher 1958, p. 84]

He noted that if evolution is to work, many adaptive mutants have to appear. Only in large numbers could mutants survive the vagaries of selection and take over the population. But adaptive mutations are just too rare for that.*

How many mutants would have to appear to ensure their survival? It's a matter of chance; there's no way to ensure their survival. We can calculate the chance that a mutation will survive if we know the selective value.

What is a typical selective value for the kind of evolution I am discussing? In the opinion of the late George Gaylord Simpson, who was generally acknowledged as the dean of evolutionists, a "frequent value" is about a tenth of a percent. He felt that a hundredth of a percent "... may be less than the average" [Simpson 1953, p. 119]. I shall therefore choose 0.1% as a typical selective value.†

---

* I was surprised that Fisher didn't make more of the point that even a mutation with a large selective value will be likely to disappear. He instead assumed that there would be many such mutations so that at least one of them would have a good chance of surviving. I think the explanation is that in 1930, when he published the first edition of his book *The Genetical Theory of Natural Selection*, no details were known of the molecular nature of the genome. Our present concept of a point mutation was then unknown, and there was no appreciation of how small the chance is of getting one.

† As Simpson [1953, p. 118] noted, a selective value of 0.1% is not easily observed, yet it is effective in the working of natural selection. He even noted that a value of 0.01% would surely be effective in natural selection, and even "much weaker selection could well be effec-

Fisher's calculations show that for only *one* mutation with a tenth of a percent selective value the odds are 500 to one against its survival. There would have to be almost 350 such mutants to have a 50% chance of survival. There would have to be more than 1100 of them to have a 90% chance.

You can now see what it will take to complete one successful step in a chain of 500. An adaptive mutation has to occur, and it has to survive to take over the population. But the chance is small that a specific copying error will appear and survive. The chance that it will appear is 1/600. For a selective value of a tenth of a percent the chance that the mutation will survive, if it appears, is 1/500. The chance that the mutation will both appear and survive to take over the population is 1/600×1/500, or one in three hundred thousand (1/300,000). That's less than the chance of flipping 18 coins and having them all come up heads.

For just a moment let's look at the chance of a species evolving into a new one if at each step there is only one potential copying error that can be adaptive. What we've found above is the chance of just one of the small steps occurring. To get a new species, 500 of them have to occur without any failures. As we shall soon see, for successful evolution the probability of each has to be very nearly one. The chance of 500 of these steps succeeding is 1/300,000 multiplied by itself 500 times. The odds against that happening are about $3.6 \times 10^{2,738}$ to one, or the chance of it happening is about 2.7×10^{-2739}. That's a very small chance! It's more than 2,000 orders of magnitude* smaller than the chance of the event I called *impossible*.

(..continued)
tive." Applied geneticists, who breed plants and animals for commercial application, by necessity deal with larger selective values, usually one percent, and even as high as ten percent. They might therefore look on a value of 0.1% as too low an estimate because it is smaller than they are used to. Experts in evolution and natural selection, however, hold that evolution must rely on smaller values.

* An *order of magnitude* is a factor of ten.

I made this last calculation assuming there was only one potential adaptive copying error through which evolution could progress at each point. At some points there may not be any; we don't know. There's no evidence that there are any. From the above we see that even if there is one, the NDT won't work. There would have to be many. But since we don't know we shall say that *maybe* there are many. If there are many, then how many must there be in the population to raise this very low chance to a level where the NDT will work? How many potential adaptive copying errors must there be to raise the chance of a successful step from 1/300,000 to 0.9727? A calculation shows that there must be about a million of them.[7]

Only if there are at least a million potential adaptive mutations will there be a chance of at least one in a million that a new species will evolve. That was the criterion we set up as the test of the theory. So we see that evolution will work only if there are at least a million potential adaptive mutations at each step.[8]

Each of the million potential mutations not only has to be adaptive, it also has to be able to add a little information to the genome. We can put this another way and say that, if a new species is to evolve from an old one, two conditions have to hold. They must apply to any stage in evolution. These conditions are:

1. An adaptation that adds information to the genome can always arise through a change in a single nucleotide:

2. At each stage of evolution there are about a million nucleotides in which a change will satisfy the first assumption.

I could have put these two conditions together and expressed them as one. But I'd rather look at them separately because they make two distinct points. Unless these conditions hold, the NDT will not work. To make the NDT work, I must assume these conditions hold. I'll call them the two *Darwinian Assumptions*. Unless they both hold, the chance of a new species evolving is too small for cumulative selection to work. As I proceed you'll see many reasons why we have to reject both assumptions. About two-and-a-half decades ago some of the most prominent evolutionists in the world were confronted with the challenge posed by the small chances of the right mutations occurring and spreading through the population. The evolutionists did not have any good answers [Moorehead and Kaplan 1967].

Let's look at these assumptions and examine some of their implications. The first assumption is that at any stage of evolution there must always be some potential nucleotide changes that can lead to a selective advantage for the organism. It also says that these changes will add a small amount of information to the genome. If adaptive mutations that add information can always occur at any stage then they should have been observed often. Have they been? Have any such mutations *ever* been observed?

There are many known cases of evolutionary change below the species level.* Scientists have seen bacteria evolve in the laboratory. There is evidence of some evolution in the field as well. For example, industrial melanism evolved in the peppered moths of Great Britain, as I described in Chapter 3. Some bacteria have mutated to become resistant to drugs that used to kill them. Some insects have become resistant to bug killers. These mutations led to new strains, but not to new species.

Why, you might ask, can't the kind of evolution shown by these examples play the role of the small steps that make up cumulative selection? Those examples have a *good* chance of

---

* There is also some evidence that *Drosophila* have formed new species in the wild [Dobzhansky 1951]. Species are defined as different if they don't mate with each other — if they *don't*, not if they *can't*. But no one has ever reported the evolution of, say, a *Drosophila* into something as different from it as a mosquito.

occurring. We know that, because they've been seen to occur often. On the other hand, I've shown you that the steps making up cumulative selection have a very *small* chance of occurring. What's the difference between these two kinds of evolution? Why can't examples like those I just mentioned play the role of the steps needed for cumulative selection?

Evolutionists don't look upon these two kinds of events as different. Yet they do differ in important ways. The steps that build up to make macroevolution must fulfill conditions that isolated steps don't have to fulfill. The conditions are:

1. They must be able to be part of a long series in which the mutation in each step is adaptive.
2. The mutations must, on the average, add a little information to the genome.

The information a mutation adds in a typical step of cumulative selection must fall within strict limits. On the average, each step must add some information. Yet it cannot be much more than one bit. Each step must add some information on the average so that information can build up over the full chain of steps that make up macroevolution. But if a mutation seems to have much more than one bit it can't be a part of cumulative selection. We would have to interpret that mutation as the switching ON of information already in the genome, as I noted in Chapter 3.

Curiously, no mutations that have selective value are known to satisfy this condition. They either reduce the information in the genome, or they seem to add too much. We'll see typical examples of this in Chapter 5. A single isolated small change, on the other hand, does not have the constraints of cumulative selection. It need not add information, and it need not form part of a series of steps that are all adaptive.

Some microevolution does not involve mutation. It instead uses the variation already in the population. The evolution of industrial melanism in the peppered moth is an example.

Both the light and dark moths have existed side by side in the population. Few potentially beneficial mutations, however, can be stored in this way. Because of the small amount of adaptive variation that can be stored in the population there cannot be a long line of adaptive gene variants sitting in the population waiting to be called into action. Therefore industrial melanism also cannot serve as a typical example of one of the steps of cumulative selection.

None of the above examples show the kind of mutations the NDT needs. In fact, there are *no known cases of evolution that meet the conditions of cumulative selection.* There are some known cases of evolution with copying errors, but they show only a kind of *micro*evolution that one cannot extend to *macro*evolution. None of them adds information. All that I know of, actually lose information. There are no known examples of copying errors that have been observed and that have been studied on the molecular level that qualify to be a step in cumulative selection. We shall therefore find we have to reject Darwinian Assumption 1, and consequently we shall have to reject the NDT.

Now let's look at the second assumption. It says the genome of a species must have a million sites where a change of a single nucleotide could yield a selective advantage. You can think of these million sites spread among some ten thousand genes, which encode some ten thousand proteins. Some of these proteins are enzymes. Some build structure, some regulate, and some deliver messages. There would then have to be something like a hundred potential adaptive changes within each of ten thousand genes.

If Assumption 2 were to hold, there would have to be a lot of freedom in the way a species can evolve. The NDT says mutations are random. The forces of selection have no say about which of the adaptive mutations will occur at any point. Selection acts only after a "good" mutation has appeared. If many such mutations occur, the selection among them is nearly random as would follow from Fisher's results.

So in the first of the 500 steps that lead to a new species there are a million choices for evolution. For each of these choices, there are in the second step another million, and so on. This continues for the full 500 steps. The process has a huge amount of freedom. If an evolutionary path were to begin a second time from the same point, the first outcome would not repeat. The odds against it repeating is a million multiplied by itself 500 times, or $10^{3,000}$, to one. By comparison, the odds against the event we called *impossible* are only 10^{45} to one. The species resulting from the second path would almost certainly be different from the first.[*]

You can compare the path of evolution of the population with a path through a huge tree-like maze. The maze is a complex of pathways that diverge, and whose every path goes through 500 nodes. Each node is a fork, and every fork has a million branches from which to make a random choice. Fig. 4.1 shows a simplified version of this tree. Instead of a million branches at each node, the figure shows only three. Imagine though that there are a million instead of three. The figure also shows only three levels of branching. But imagine the tree going on for 500 levels. As the population moves through the maze it comes to a fork, or branch point, at each of its 500 steps where it has a million branches to choose from.

There's a separate exit from the maze for each path that might be taken. Each exit from the maze corresponds to a possible form for a typical member of the population. The

---

[*]My son Daniel pointed out to me that the odds may well be less than this, because many of the million possibilities at different steps will be the same. If I were to take the extreme case, where *all* of the million possibilities were identical at each of the 500 steps, then the figure I gave would be too large by a factor of 500!, which is about $10^{1,134}$. I must therefore reduce the number $10^{3,000}$ by this factor. The resulting odds are then $10^{1,866}$, still a large margin from the odds of 10^{45} to one.

number of exits is the number of alternate forms a population could take if it should evolve under Assumption 2.

The independent evolution of the same trait in different lines of descent is called *convergent* evolution. If two lines start together and if the evolution in both is influenced by the same factors then the evolution is called *parallel* evolution. The development of the wing of the bird and the wing of the bat are said by evolutionists to be *convergent*. The develop-

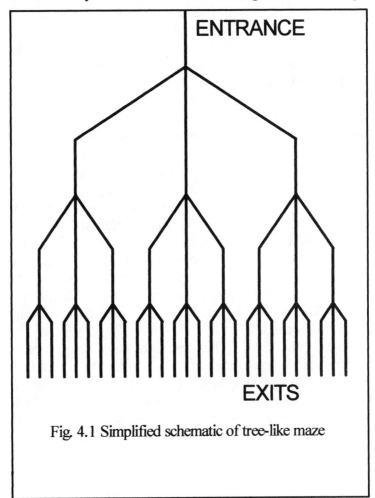

Fig. 4.1 Simplified schematic of tree-like maze

ment of the convolutions of the brain in the horse and in the ape are *parallel*. Both convergent and parallel evolution are different sides of the same coin, and both could really be called *convergent*.

From the maze I've just described, you can see there's a lot of freedom in how the population goes through it. At every node there's supposed to be a random choice among the million ways to go. If the choices are really random, then the chance of the same trait evolving twice is so small that we should have to call it *impossible*. To get convergence, two lines of descent would have to make the same random choices at each of the 500 nodes they meet so that they come out at the same place.

The odds against the population taking exactly the same path twice is $10^{3,000}$ to one. But convergences are almost never perfect. The bat wing is not built exactly like the bird wing. So let's say that, in a convergent transition to a new species, not all the 500 choices had to be the same. If only 100 of them had to be the same, the odds against it would be 10^{600} to one. Those odds are still enormously greater than the odds against the event we called *impossible*. And don't forget, I'm dealing here only with the transition from one species to the next. No evolutionist would try to say a complicated organ, like a wing or a kidney — to say nothing of the eye — developed in a *single* species transition. For the evolution of a complex organ we would have to allow many thousands of steps.

We see then that, if the variation arises from random copying errors, convergent evolution is impossible. But convergence plays an important role in evolution. Evolutionists call on it whenever they can't explain how two similar traits came about by descent. Every one of the many instances where they must invoke convergence is a contradiction to the NDT.

———

We find convergence throughout the whole of the taxonomic system. The late George Gaylord Simpson noted that convergence is found at all levels [Simpson 1953]. Striking examples of convergence are even found between members of different phyla. Arthur Willey [1911] wrote a whole book on convergence in evolution. There are thousands of examples of convergence that I could give, but I shall limit myself to just a few.

One amazing example of convergence is the ultrasonic echolocation systems (like sonar) found in animals scattered through the vertebrate phylum. Their sophistication rivals the most advanced military systems. Bats have an echolocation system, and so do toothed whales and dolphins. The system is also found in some birds. According to the experts, these systems could not all have been derived from a common ancestor. They all agree that these systems evolved in each of these groups *independently*.

South-American electric fish and African electric fish both "see" in dark murky water by measuring the distortion of the electrostatic fields they generate in the water around them. These two groups of fish are believed to have developed their electrostatic-imaging systems independently

Some fish kill their prey by electrocution. High-voltage generators for killing prey are found in electric eels, in torpedo rays, sting rays and electric catfish [Nichols and Hubbs 1967]. These electric generators are held to have evolved independently in the eels, the rays and the catfish [Hubbs 1967].

The most amazing example of convergence is the visual system. Systems for sensing light are found in many animal phyla. The simplest of them is the eye spot of the one-celled *Euglena*, which is not even in the animal kingdom. The most complex is the eye of the vertebrates, including that in humans.

In three animal phyla there are image-forming eyes. One of these is the vertebrates, of which the human eye is a typical example. If this eye developed according to the NDT, then it must have appeared in the first vertebrates, which are estimated to have arisen some half a billion years ago.* The *second* is the mollusc phylum, some of whose members have eyes much like those of the vertebrates. In this phylum, the octopus, the squid, and the cuttlefish have such eyes. In the arthropod phylum, the insects, spiders, and crabs have image-forming eyes.. Until two years ago, biologists unanimously held that eyes evolved independently as many as three or four dozen times [Zuker 1994].

Just two years ago, a striking identity was reported between an insect gene and a vertebrate gene [Quiring et al. 1994]. This gene has been found to control eye development both in insects and in vertebrates, including humans. The genes in these two different phyla are 94% identical. This new finding makes convergence look so improbable that, even without making any probability calculations, the authors suggest that

> the traditional view that the vertebrate eye and the compound eye of insects evolved independently has to be reconsidered.

But maybe the eyes neither converged nor developed independently. There's a third option, which I shall take up in Chapter 7.

Exceeding the eye in complexity is the human brain. It's probably the most complex object in the universe [Fischbach 1992]. It is said to have evolved from the much simpler brain of the primitive fishes. The brains of fish, amphibians, and reptiles are all smaller[†] and their cortices[‡] are smoother than those of mammals. Even among the mammals, the brains vary greatly. We ascribe greater intelligence to those animals

---

* Because all vertebrates have the same kind of eye, the eye must have appeared before the vertebrates branched into their separate classes.
[†] Not the absolute size of the brain, but its size relative to the size of the animal is conventionally taken as a measure of intelligence.
[‡] Plural of cortex. The cortex of the brain is the outer layer of its two hemispheres.

with the larger and more convoluted brain. Of modern mammals, the brain surface goes from the small smooth brain of the anteater to the large and highly convoluted brain of the horse, the whale, the elephant, and, of course, man. From studies of fossil skulls, paleontologists have concluded that the mammalian brain *has evolved in parallel* in the different orders, and even families, of mammals. Evolutionists hold that the most complex object in the universe has developed from random variations several times!

The list of convergences goes on and on. Convergences occur throughout the plant and animal kingdoms. Whenever evolutionists invoke convergence, they have to reject *Darwinian Assumption 2*, which says that, in the genome of any organism, there are always a great many nucleotides in which a change can be adaptive. Since the NDT needs both assumptions 1 and 2, to invoke convergence is to refute the NDT.

The average person finds it hard to believe that complexity and sophistication of such high order was developed by having natural selection organize random events. Evolutionists try to teach them that they must alter their thinking to be able to accept such an incredible conclusion [Dawkins 1986]. As we have seen, and as shall see further, the average person's intuition is correct and the neo-Darwinists have gone awry in their sophistry.

———

I have shown my calculation to evolutionists and the only objection they have come up with is that my calculation was about strings of DNA symbols in the genome (the genotype), and we see convergent evolution in the phenotype. They say that there might be *many* genotypes that can lead to the same phenotype. So maybe my maze should have fewer paths.

My answer has been the following. There may, indeed, be many genotypes that lead to the same phenotype, and maybe I do have to reduce the number of paths in my maze. We don't know so much about convergence in the genotype as we know about it in the phenotype. But we do know that the genotype determines the phenotype. Therefore freedom in the genotype means freedom in the phenotype as well.

The freedom in the phenotype may not be so large as the freedom in the genotype, but still the number of possibilities is large. Let the maze be built on the basis of the phenotype — it will still have an enormous number of paths. The maze for the phenotype may have fewer branches at each node than the maze for the genotype. There may be less than a million — maybe only ten thousand. In that case there would be $10^{2,000}$ branches. The odds against coming out the same place twice would still be enormously larger than the odds against what we called the *impossible event*. Since we would still have to call it *impossible*, we have to rule out phenotypic, as well as genotypic, convergence.

Since the genetic discovery reported by Quiring and colleagues [1994] we see that phenotypic "convergence" *does* imply genetic "convergence." Convergence in the genotype may very well be just as striking as convergence in the phenotype.

Nearly all our examples of convergence have been of the phenotype because that's all we could observe up until the last twenty years or so. Now, however, you see we have the tools that allow us to peer into the genome. I wouldn't be surprised to find more cases like that of the gene for eye development.

There is another recently found example of convergence in the genotype. It's a case of convergence that has been studied at the molecular level, and the results of that study relate directly to the genome. A team in Allan Wilson's lab at Berkeley has compared versions of the same enzyme found in several mammals [Stewart et al. 1987]. The enzyme they studied is called *lysozyme*, and the mammals they focused on were the cow and the langur monkey.

Lysozyme is an enzyme found in the stomach of ruminants. The ruminants are mammals that have hoofs and chew their cud. They include the cow, sheep, goat, giraffe, and deer.

The stomach of the langur monkey, which is a primate and not a ruminant, is similar to that of the ruminants. Neo-Darwinians say the evolution of these animals must have converged to produce such similar stomachs.

No one thinks the langur descended from the ruminants, nor does anyone think the langur monkey and the cow have a common ancestor that had such a stomach. Their common ancestor had to be some primitive little mammal that lived way back when the dinosaurs ruled the earth. No one thinks there were any ruminant stomachs so early as that. All hold that the evolution of this trait in the two animals converged because they can't explain it by descent.

The langur and the ruminants are classified far apart within the mammals, as shown in Fig. 4.2. The ruminants and the langur are in different orders. One is an artiodactyl, the other is a primate. Because their stomachs are alike, the experts hold they must have converged.

In the forward part of the stomach of both the langur and the ruminant there is a special fermentation chamber. Both types of animals use bacteria to break down cellulose. Cellulose is the main component of the leaves and grass in the animal's diet. There is food value in the cellulose, but the animal cannot digest it on its own. It doesn't have the right enzymes. It hands that job over to bacteria, which it hosts in the fermentation chamber. The bacteria do have the right enzymes. The animal lets them eat and digest the cellulose in the leaves and the grass it eats. Then the animal transfers the bacteria to a rear compartment of the stomach and digests them. Lysozyme breaks down the bacterial cell wall, and allows the animal to digest the nutrients in the bacterium.

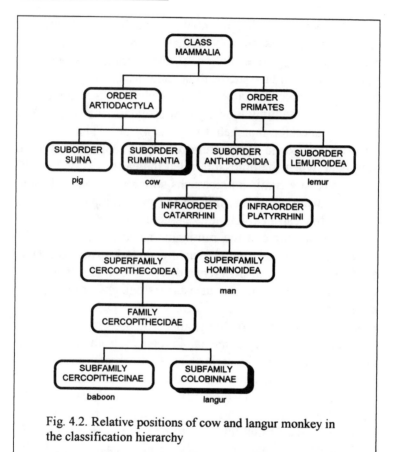

Fig. 4.2. Relative positions of cow and langur monkey in the classification hierarchy

Other mammals also have lysozyme, but they have much less of it. They use their lysozyme only to kill harmful bacteria.

The Berkeley group analyzed the sequences of amino acids of the stomach lysozyme of man, baboon, rat, horse, cow, and langur. From their data they concluded that seven amino acids have converged in the cow and the langur [Stewart et al. 1987]. The convergence of the lysozyme is in addition to the convergence in the anatomy of the stomachs.

With the genetic code, those amino acids can be translated back into DNA nucleotides to within the redundancy of the code. It turns out that no less than nine nucleotides had to mutate to change those seven amino acids.* If the cow and langur lysozymes really did converge they would have had to do so in no less than nine steps. It could have taken *more*, but it could *not* have taken *less*. The change of each nucleotide would take a *minimum* of one step. Each of those nine changes had to have a selective advantage. Each also had to be lucky enough to avoid a chance extinction and to spread through the population.

The second Darwinian Assumption said that, at each step, any of a million potential mutations could have been selected. My example of the horse is typical of animals. If the NDT is to work, the same has to hold for other animals as well. If the lysozyme of the langur and the cow really did converge, they would have had to take the same path through the maze. At each of the nine steps the same choice of one out of a million had to be taken in both lines. Notice that now we're dealing with DNA nucleotides and not with phenotypes.

Under the second Darwinian assumption then, the chance is very small of getting the same nine steps in the evolution of the langur and the cow. The chance that the first adaptive mutation in the lysozyme gene would be the same in both lines is one in a million.† The chance of getting all nine of them the same is one in $1,000,000^9$, or one in 10^{54}. This chance is a *billion* times smaller than the chance of the event I called *impossible*. Note that this chance takes account of the convergence of only the lysozyme molecule. The chance of convergence of the stomachs ought to be *even smaller*, and it

---

* If the lysozyme genes of these species could be found on the genome and their base sequences determined, we would know exactly how many nucleotides had to mutate for this convergence. It could be *more* than nine, but nine is the minimum.

† This chance derives from the million branches that we calculated when we considered convergence in the genotype. Since we are dealing with the genotype here, we don't have to reduce this number as we suggested earlier in this section, to account for the possibility that many genotypes may determine the same phenotype.

has to multiply the chance of $1/10^{54}$ to give us the chance of both the enzymes *and* the stomachs converging.

The NDT then says it's impossible for these two animals to converge like this. Yet there they are. How would the NDT explain it? The contradictions in the theory caused by the need to invoke the Darwinian assumptions compel us to reject the theory.

———

Some evolutionists have suggested that evolution may not have to wait for the right mutation to occur just when it is needed. A useful one might already be stored in the population. But this would mean that a huge number of useless mutations would have to be stored to have one that would be useful.

Some have noted that adaptive DNA sequences could be stored in the population more efficiently than by having just one in each individual. One might like to say that a recombination can sometimes yield a base sequence that is adaptive in some environment. Many more *potential* base sequences could be stored this way than can *actual* ones. As I noted in Chapter 3, Ayala estimated that there could be as many as $10^{2,017}$ potential recombinations. Could this large number of potential recombinations supply the variation for selection to work?

How many of these potential recombinations would have to be adaptive to make this idea work? Let's take a population of a million, and let each individual on the average have 10 offspring. Then in a million generations, there would be a total of 10 trillion (10^{13}) reproductions. In each of these reproductions there is the possibility of a recombination. Let's say for simplicity that there is an average of one recombination in each reproduction. For evolution to proceed one step in this period, one of these 10^{13} recombinations would have to have a positive selective value; or, as we are saying, one of them would have to be *adaptive*.

To have a good chance of getting *one* adaptive recombination in this period, a lot of the potential recombinations would have to be adaptive. To have a chance of at least one in a million of getting one adaptive recombination in 10 trillion replications, there would have to be $10^{2,017}/10^{19}$ adaptive ones. That means that $10^{1,998}$ potential recombinations would have to be adaptive.

With this number of adaptive possibilities, there would be a one-in-a-million chance that one of them will appear in the population during the million generations. Actually, to get this chance of an adaptive recombination that will *survive* in the population, we need somewhat more than $10^{1,998}$. But never mind. That number is already too big to allow convergence.

Now instead of a number like a million that we found in the calculation of horse evolution, we have this super astronomical number. If there can be no convergence when there are a million potential adaptive mutations, there surely can be none when there are $10^{1,998}$ of them.

With this we see that copying errors, even if they are stored as potential recombinations, cannot provide the random variation needed by the NDT.

———

I hope I have shown you here why the NDT doesn't work. I have shown it through an example. Fred Hoyle, astronomer, mathematician, Fellow of the Royal Society, and retired professor at Cambridge University, together with Chandra Wickramasinghe, chairman of the Department of Astronomy and Applied Mathematics of the University of Cardiff, have arrived at the same result in a more general mathematical way. They have presented a mathematical disproof of the NDT in a small book of only 34 pages, entitled *Why Neo-Darwinism Does Not Work*. They call what they have done a "simple and decisive disproof of the 'Darwinian' theory." [Hoyle and Wickramasinghe 1982].

———

I've shown in this chapter that the NDT has not delivered on its promise. Darwin's theory, when he proposed it, seemed to refute the Argument from Design. The theory was supposed to show that life could develop from a simple beginning to the complex organisms of today. It was supposed to be able to do this without the need for a designer or a creator. It was supposed to substitute *chance* for *design*.

The NDT's claim is the same as Darwin's. It claims to explain how all life evolved from some simple beginning. It claims to explain how all the complexity of life evolved in a natural way. It claims to explain evolution as a process that occurs through a combination of chance and the known laws of nature. It, too, claims to have substituted chance for design.

I have shown so far that on theoretical grounds random mutations cannot form the basis of evolution. The information of life could not have been built up the way the NDT says it was. Evolutionists have not succeeded in finding a random source of the variation that will make the NDT work. In the 1940's, when we knew almost nothing of molecular biology, evolutionists were satisfied that the NDT could explain evolution. In the 1990's we know too much about molecular biology to be satisfied.

Genetic rearrangements don't work because they can't build up the necessary information. Moreover they don't seem to be random. Copying errors don't work because the chance is too small that they could build up the necessary information. The only way to make copying errors work is to declare convergence impossible. But convergence is too pervasive in the living world to permit that option. If copying errors are the variation of the NDT, then the theory predicts the important events of evolution to be nearly impossible. If a theory predicts events to be *nearly impossible* then one cannot justifiably say that it *explains* those events. If the NDT cannot explain what it claims to be the most important events of evolution, we must reject it.

There is nothing new in what I am presenting. Some biologists who have thought about the problem of randomness have also concluded that random variation cannot form the basis of evolution. After I came to this conclusion, I found that Ho and Saunders [1979] had arrived at it through a different path. Many biologists are skeptical about the power of small mutations to accumulate to achieve macroevolution. In this chapter I have shown you why one should be skeptical. This chapter forms the basis of a serious challenge to the NDT.

———

NOTES TO CHAPTER 4

1. **Random variation.** Let me emphasize here that I am not asking if *evolution* is random. I am asking if the *variation* that forms one of the two pillars of the NDT is random. I make this point because I want to avoid the confusion that might arise if you try comparing my discussion here with what evolutionists often plead [e.g. Dawkins 1986]. (For an interesting discussion of how one might tell if evolution is random, see Katz [1987]). One can indeed get a nonrandom result from random variation if one applies selection to the variation. Evolutionists think of evolution as doing just that. An extreme, but trivial, example is that I can create a message of my choice by taking or rejecting alphabetic characters that are generated randomly. Since the message is of my choosing, it is not random. My discussion here centers instead on the important question of whether or not *random variation*, even with natural selection, can lead to evolution.

2. **Mutation rates**. Forty and fifty years ago scientists thought that mutation rate was not a limiting factor in evolution. Mutation rates were quoted to be typically between 10^{-5} and 10^{-6}, and these were felt to be fast enough for evolution [Simpson 1953]. These rates are not the numbers I have cited.

Both sets of numbers are right. There is no contradiction between them because they refer to different things. What were called *mutation rates* 50 years ago are not the same as what I am here calling *mutation rates*. *Mutation rates* then were per *gene* or per *battery of genes*. My *mutation rates* are per *nucleotide*. Forty and fifty years ago we didn't know the structure of DNA. Mutations were known only by their effects on the phenotype, and these effects were almost always damaging. A mutation in any one of many bases in a gene could disable a protein. Changes in any one of many different bases of a gene can have the same effect in the phenotype and all were considered the *same* mutation.

The mutation rates Simpson cited were the rates at which a gene, or a battery of genes, could be disabled. What I am calling mutation rates are the rates at which a specific nucleotide is changed.

3. **Number of births per evolutionary step.** Here's how I estimated the 50 million births per step in the evolution of the horse. I got most of my data from the late George Gaylord Simpson, who was a well-known paleontologist, and an authority on horse evolution. He estimated that the whole of horse evolution took about 65 million years. He estimated there were about 1.5 trillion births in the horse line.

How many of these could we assign to one step of evolution? The experts say the modern horse has evolved through some 10 to 15 genera. If we say the horse line, from *Hyracotherium* to the modern horse, went through about five species in each genus, then the horse line with its 1.5 trillion births went through about 60 species. (That's about a million years per species, and is consistent with other, independent, estimates I've seen.) That would make about 25 billion births per species. If I divide 25 billion births per species by the 500 steps per species transition, I get 50 million births per step.

4. **We neglect the chance of multiple copying errors.** We shall limit ourselves to getting only *one* copying error at a time and neglect the chance of getting *more* than one. We

can do this because the chance of getting two or more *specific* mutations at once is small. For animals, we've seen that the mean mutation rate per nucleotide is about one per ten billion per birth. The mean rate of mutation of a specific *set of two* nucleotides per birth is less than that of a single nucleotide by a factor of ten billion, or 10^{10}. The rate for a specific *set of three* is less by a factor of 10^{20}, and so on. We have found that the chance of getting a single mutation in one step is 1/600. The chance of getting a double one is one ten billionth of that, or 1.7×10^{-13}. The chance of getting a triple is one ten billionth of *that*, or 1.7×10^{-23}. The chance of getting higher-order mutations would be smaller still. We are therefore justified in neglecting double-order and higher-order mutations.

5. **There have to be a million potential adaptive mutations to make the theory work.** Actually the number turns out to be about 1,080,000. We can check this by verifying that at least one out of 1,080,000 possibilities will occur with probability 0.9727. We found that the chance for a particular mutation to occur and take over the population in one step is one in 300,000, or a probability of 0.000,003,333. The chance that it will *not* occur is one minus this number, or 0.999,996,667. The chance that *none* of the 1,080,000 potential mutations will occur and take over is $0.999,996,667^{1,080,000}$. This works out to be 0.0273. The chance that at least one of the potential adaptive mutations will occur and survive is one minus this number, or 0.9727.

6. **A different choice of parameters cannot make the theory look better.** Some might think that by changing the choice of some of the parameters I used I might yield an answer more favorable to the theory, but that is not so. To make the theory work I would have to raise my parameter values enough to gain a factor of a million. I chose values for four parameters. The first was a typical value for the mutation rate, the second was the number of births per evolutionary

step, the third was the number of steps in a species transition, and the fourth was a typical selective value of the mutations.

I can't increase my value for the mutation rate. I was careful to choose for it the (geometric) mean value of observed rates. I have already noted that a much higher mutation rate would destroy the species. Moreover, from an evolutionary point of view the observed rates must be nearly the optimum ones, and the values I chose are therefore the ones my calculations must take account of.

Nor can I raise my values for the number of births per evolutionary step, nor for the number of steps in a species transition, nor for the selective value. I got these from estimates made by a paleontologist and a geneticist, both of whom are acknowledged to have been experts in their fields and in evolution. Both men are also well known as having been strong advocates for the NDT. Moreover, I don't know of any prominent evolutionist who disagrees with these estimates. If I should raise my values of any of these parameters, I should be forgoing typical values to take on unrealistic ones. Furthermore, if the NDT is to work, it has to do so with *typical* values such as those I have chosen. So there is little room for any increase in my choice of these parameters.

Chapter 5

CAN RANDOM VARIATION BUILD
INFORMATION?

THE company's four vice presidents sat before the CEO in his grossly oversized office. They had all played a major role in the company's phenomenal growth in the past four years. Each of them became a multimillionaire when the company went public. The CEO got up from his chair and looked through the wide picture window out onto the boats floating lazily on the Charles River.

"If there weren't a law against insider trading," he was saying, "we could each make millions tomorrow. Our announcement of the new product ought to drive our stock up at least ten points."

The intercom buzzed. He spun around, poked it, bent his ear to it, and barked, "Yes!"

"The professor from MIT is here," came the voice from the box on his desk. "He's the one who's had an appointment with you for over a month. He wants to tell you about a new design, or something like that. He's in the waiting room outside. Shall I tell him you can't be disturbed?"

"I really can't afford the time to see him. But he was a colleague of my father's and I've put him off several times already. — You know what, we're actually finished here. Send him in."

"Shall we go, Bob?" asked the vice president in charge of production.

"No. This won't take long. Maybe he has something interesting to tell us that you ought to hear."

125

"Hello," said the newcomer as he entered. Glancing at the CEO and then at the four men seated before him he said, "I hope I'm not intruding."

"No, no. We just finished our meeting."

"I'm Pat Reynolds from MIT."

"Yes, I know who you are. My father used to mention your name when the two of you were teaching together at Purdue. I understand you want to tell me about some product design. But for my protection and for yours, I can't allow myself to discuss it with you unless you have some patent protection."

"Yes. Well that's OK. The Institute has already applied for a patent."

"OK, then, go ahead."

"It's about an improved design for your latest product. Actually, one of my students came up with the idea in a term problem. He analyzed your product and discovered that it could be redesigned using only a tenth of the components that are now in it. He didn't reverse engineer your box, but he made his own design from scratch. I thought he had a good idea, so I worked with him 'till we had what we think is an optimum design. I think you ought to be interested because our design ought to reduce your costs by about a factor of ten."

"Look, Pat — may I call you Pat?"

"Sure. My graduate students do."

"Pat, I want to thank you for your interest in helping us. But it doesn't pay us to invest any more money into the design of that product. Our marketing department estimates the remaining market life of that product to be only another year. Ours is a rapidly changing business. Old products die fast, and we constantly have to come up with new ones. I'm sorry, but there's just no point in me looking at your design."

"I see. Well, then, perhaps I can leave my student's resume with you. He'll be getting his degree this June and he'll be looking for a job. He's brilliant, and I think he could do you a lot of good in your future designs."

126

"OK, fine. Just leave the resume with the receptionist on your way out. And thanks for thinking of us."

The professor snapped his briefcase shut, stood up to his full height, and headed toward the door. At the door he turned and said with a smile, "Please give my regards to your father."

"Sure will," was the answer with a return smile.

As the door shut, the CEO grinned to the men still seated before him.

"You see," he said, "this is why we have no competition. Everyone assumes we design our products from scratch. No one suspects our real secret, that we buy up obsolete computers and accessories for next to nothing and then figure out how we can make a marketable product out of them.

"We know very well what Pat's student discovered — that we have ten times more components in that product than we need. Most of them aren't being used. But they didn't cost us anything, so what's the difference? The most expensive part of that product is the plastic case around it, because we had to have it made specially. Everything else in it is junk. Gentlemen, let's face it. We're in the junk business.

"The name of the game is adapting to the marketplace, not elegant design. Our products sell because the price is low. We pick up a few million obsolete computers and look for ways to adapt the parts to make a salable product. We usually disable nine tenths of the components in a card to get a marketable function. We're not at the cutting edge of technology, gentlemen, but we're making a fortune!"

———

In living organisms, adaptation often takes place by reducing the information in the genome, much like the way the company in the story above makes a useful product out of junk. The name of the game is adapting to the environment, not necessarily increasing the information in the organisms. Yet, the neo-Darwinian theory (NDT) has captured the minds of the Western World not because it explains a few adaptations here and there, but because it pretends to account for the appearance of all the complexity in life. It pretends to account for how eyes came to be, how brains came to be. Is there any evidence to justify this pretense? Is there any evidence that evolution can build up information in living things? Or can it only reduce or destroy information?

In the last chapter we examined a frequently cited example of macroevolution that the NDT is said to explain. We saw that the theory can't explain it. In this chapter we shall examine several examples to see if there has ever been a clear case where evolution has added information.

———

No one knows of an inherent *drive* in living things toward greater complexity, yet if the life of today had evolved from some simple form, then evolution must have built up a lot of information and complexity. Today's life is very much more complex than whatever simple form of life is supposed to have sprung up from inorganic molecules some four billion years ago. If life as we see it today has evolved, then it must have grown in complexity. Any theory that is supposed to explain Darwin's concept of descent must explain the growth of information and complexity.

Most experts hold that evolution as described by the NDT is the mechanism that built up the complexity now found in all living things. John Bonner of Princeton University wrote [Bonner 1988]:

> It is evident that there has been, over the course of time, a continuous increase in the number of species, and an increase in the size and complexity of organisms ... [p. 226].
> ... evolution usually progresses by increases in complexity [p. 228] ...

If land vertebrates have evolved from a fish, then evolution has built up information and complexity, and led to a new form of life with new specific structures and functions adapted to the dry-land environment. The land animal has structures and functions the fish doesn't have. Some say that evolution may not have added net complexity in that transition, because a fish has some structures and functions the land animal doesn't have [Hinegardner and Engelberg 1983, McCoy 1977]. A dog may not be any more complex than a salmon, it's just different.

In evolving to a land animal, the fish lost some of the complexity it no longer needed, and picked up other complexity that would help it live on land. There may or may not have been a net gain of complexity. If evolution has converted a fish's fin into a forearm, it replaced one form of complexity with another. Whether or not it produced a net gain in complexity is not the point here. The point is that new information must have been added. That's the way evolution is supposed to have worked in the long run.

In some special cases an organism might adapt simply by dropping some obsolete complexity without adding anything new. This would be a case only of losing information without gaining any. If evolution only worked by losing information in this way, it could not have led to the development of the complexity of life we see around us. The vertebrate eye or its immune system, for example, could not have evolved by loss alone.

If evolution worked only by losing information, how could it have originally built up the information that it loses? How could it have built the eye and the system of vision in the first place?

- How, for example, could it have developed the photoreceptor cell of the eye, which is so sensitive it can detect a single photon [Schnapf and Baylor 1987]?

- How could it have developed the pigments of the retina, which make color vision possible [Levine and MacNichol 1982, Rushton 1962]?

- How could it have developed the muscular system of the eye through which the brain exercises precise control over eye movements [Bahill and Stark 1979]?

- How could it have developed the visual cortex, which is the part of the brain that receives the retinal information and interprets it, enabling, among other things, binocular vision [Pettigrew 1972, Hubel 1963]?

- How could it have developed the extraordinary ability of the eye-brain system to correct for optical distortions [Kohler 1962]?

- How could it have developed that remarkable computer in the retina which processes and compresses the information before sending it to the visual cortex [Michael 1969]?

Evolutionists have not suggested that any of these intricate and sophisticated systems evolved in one jump. They say they came about gradually through a long sequence of small steps. The information defining them is supposed to have built up step by step. On the average, each step must then have added a little information. Some steps may not add anything, but most of them must add something, and on the average they must each add a little.

McCoy [1977] has cited a work by Joachim Barrande who, in 1871, tried to disprove Darwinian evolution. Barrande noted from his studies of the evolution of Trilobites that one cannot

> ... recognize the slightest gradual and constant progress in their apparent organization throughout the immense duration of the entire tribe ....

Barrande considered this to be evidence against a theory that

> ... animal evolution would have to have taken place in an order determined by the successive degrees of organization, proceeding from the simple to the complex ....

McCoy declared Barrande's proof invalid ("the oddest" of "the many curious 'disproofs'") because he assumed "Darwinian evolution required a continual increase in the complexity of life forms throughout the geological record." Barrande might have had a valid argument in mind. He just didn't use the right words or give modern examples.

I showed in Chapter 4 that describing horse evolution according to the NDT leads to a contradiction. In this chapter I'll show that among all the mutations that have been studied, there aren't any known, clear, examples of a mutation that has added information. Of course, I cannot, nor can anyone else, demonstrate the negative thesis that there are no such mutations at all. Surely, I don't know *all* examples of mutations. There surely are examples out there that nobody has yet discovered. But so far as is known, or at least so far as I know, there are no such examples.

Daniel McShea of the University of Michigan compared the vertebral columns of several lines of mammals. He defined complexity in terms of the changes in shape from one vertebra to the next. He concluded that complexity increased in some lines and decreased in others [McShea 1991, 1993]. George Boyajian of the University of Pennsylvania and Tim Lutz of West Chester University studied the juncture between the shells of extinct mollusks. They defined complexity of these junctures in terms of how irregular their pattern is. They found there were increases in complexity, even though the trend is not uniform [Boyajian and Lutz 1992].

But in neither of these cases is the gain in complexity the clear result of random variation. As I shall discuss in Chapter 7, changes in bones and shells can be caused directly by the environment, and they may not at all be the result of a mutation, random or not.

In this chapter I'll bring several examples of evolution, particularly mutations, and show that information is not increased. I don't say it's impossible for a mutation to add a little information. It's just highly improbable on theoretical grounds. But in all the reading I've done in the life-sciences literature, I've never found a mutation that added information. The NDT says not only that such mutations must occur, they must also be probable enough for a long sequence of them to lead to macroevolution. There is one example of a chain of three mutations in bacteria that I, at first, thought demonstrates successive increases of information. But further examination of these mutations showed that none of them added information — all of them lost it. I shall discuss this example in some detail later.

That reminds me of the story my brother Kenny tells of the guest speaker at a dinner of leading industrialists who was introduced as a successful speculator who, in two weeks, made 370 million dollars in uranium. After the speaker had risen to speak and acknowledged the applause, he said that since he was a stickler for details, he wanted first to correct a few in the introduction.

"First of all," he said, "it wasn't uranium, it was uranium oxide.

"Second, it wasn't two weeks, it was 10 days.

"Third, it wasn't 370 million, it was 730 million.

"And fourth — I didn't make it, I lost it!"

Just like a fortune can't be made by losing money, evolution can't build up information by losing it. Moreover, before you can lose money, or information, you first have to make it.

———

Robert Williams of the University of Tennessee and his colleagues at the Universidad Autønoma of Madrid, reported an example of rapid evolution in the cat brain [Williams et al. 1993]. They have been able to compare directly the domestic tabby cat (species *catus*) with its evolutionary ancestor of 3,000 years ago. They are able to do this because they found a species of cat that has descended from the same ancestor, and has not changed much since. This is the Spanish wildcat (species *sylvestris*).

Williams, an expert in neurophysiology, compared the brains, the optic nerves, and the retinas of the two cat species. He found the tabby cat to have only two thirds the brain cells of the wildcat, and about two thirds the nerve cells in the optic nerve. The tabby cat has about half the number of ganglion cells and about 40 per cent of the maximum cone density in its retina than has the wildcat. Evolution in this case, has led not to an *increase* in complexity, but to a *decrease*.

In its early stages, the tabby cat embryo develops many more cells in its optic nerve than are found in the adult. As the embryo grows, many of these cells die. Williams and his colleagues found that the wildcat embryo has the same number of cells in its optic nerve as does the tabby embryo. But in the tabby cat, more of the cells die than in the wildcat. Evolution in this case, then, occurred by cell death in the embryo.

We see here how information, and therefore complexity, has been lost in this step of evolution. Although there is no evidence that the tabby cat evolved from the wildcat by a random mutation, a mutation that would cause such a change does not seem, intuitively, to be very improbable. Williams and his team noted, though, that the mechanism of cell death that drove this step of evolution cannot be typical of the steps that had to occur to produce the long-term increase in the brain size of mammals shown in the fossil record.

———

As clear as the case of cat evolution seems to be, we still cannot say just how that evolution came about. Did it occur through random variation and natural selection, as the NDT describes? We can only guess. We don't know for sure.

Let's then turn to clear cases of random mutations to see if we can find any evidence of neo-Darwinian evolution that might build information. For a mutation to play a role in evolution it has to be of some benefit to the organism, and it has to be selected. Are there any mutations like that?

Yes there are. There are some mutations that are known to benefit an organism in special cases. We have already seen that a back mutation can restore a lost function. Such a mutation will benefit the organism. There are other examples of "good", or *adaptive*, mutations. There are those that give bacteria a resistance to streptomycin. There are those that give insects resistance to DDT. Also, applied geneticists have shown they can manipulate some quantitative traits, such as the size of tomatoes, the amount of milk a cow will give, the amount of protein a wheat grain will yield, and the number of eggs a hen will lay. In breeding experiments carried out over many years, geneticists have shown they can get large increases in these traits.

Can any of these examples serve as a prototype of the small adaptive mutations that could be strung together in a long series to get macrocvolution? Let's look at some of these examples of mutations and see how much information gets into the genome. Let's also try to see if any of them can be a model for the mutations the neo-Darwinian theory needs. Can any of them serve as the small adaptive mutations that could accumulate into a long series to make a macroevolutionary change? Let us address this question.

―――――

Before we enter the subject of mutations and information, let's first see how *information* is related to *specificity*. The more specific a gene, the more information it contains. In general, the more specific any message, the more information it contains. The information in a gene is the same as the information in the protein it encodes.

To see the relation between information and specificity, let's take an example. Let's look at the information in numbers that could be used to denote rooms in an apartment building. Let's think of an apartment building with 16 apartments, four apartments on each of four floors, as shown in part *a* of Fig. 5.1. The floor plans of all floors are identical and are as shown in part *b* of Fig. 5.1. Each apartment has four rooms. Altogether there are 64 rooms in the building, and we shall number them from 0 to 63. (It's convenient to start the numbering from 0 rather than from 1.)

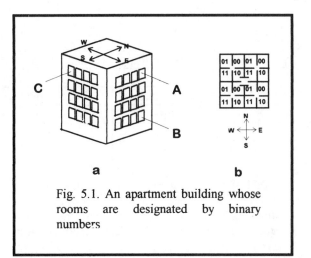

Fig. 5.1. An apartment building whose rooms are designated by binary numbers

I'll write each room number in binary form. I need six binary digits to count from 0 to 63. Table 5.1 shows how the six-digits are assigned to each room. Note in the table that the first digit of each room number tells us if the room is in the upper or lower half of the building. The second digit tells us if the room is on an even or an odd floor. The functions of the rest of the digits of each room number are as shown in the table. The fifth and sixth digits of the room number are shown in part *b* of Fig. 5.1 for each room on a typical floor. These pairs of digits are, of course, the same for all floors. For example, the room in the northeast corner of the fourth floor, which is the one whose window is marked "A" in the figure, has the number 110000. Fig. 5.2 shows the interpretation of each of the six binary digits in the number that specifies Room "A". This interpretation of the digits follows the specification defined in Table 5.1. The decimal form of that number specifying Room "A" is 48. Room number 0 (000000) is the one on the first floor in the northeast corner of the building, whose window is marked "B". Room number 63 (111111) is the one in the southwest corner on the fourth floor whose window is marked "C".

DIGIT	0	1
1	Lower half of building	Upper half of building
2	Odd floor	Even floor
3	North apartment	South apartment
4	East apartment	West apartment
5	North room	South room
6	East room	West room

Table 5.1. Binary representation of the numbers of the rooms in the apartment building of Fig. 5.1. For each digit (left-hand column) the entries show, according to the location of the room, if it is a 0 or a 1.

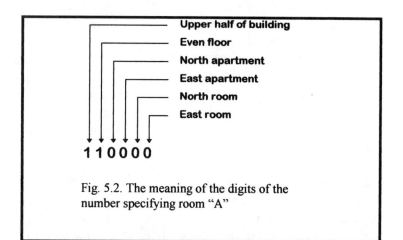

Fig. 5.2. The meaning of the digits of the number specifying room "A"

The number denoting a room serves as the address of the room. It is specific because it specifies a room without ambiguity. It carries six bits of information — one bit for each of the six digits. Now let me write an address as 10XXXX, where I mean the X's to stand for either 0 or 1, without telling which. The X's are ambiguous. The number denotes a room on the third floor without being any more specific than that. It could be any room in any apartment on the third floor. What I wrote carries only 2 bits of information, which are the

1 and the 0. The X's don't carry any information — only the 1 and the 0 do. Similarly, the address XX1010 carries 4 bits of information and denotes a room in the southeast corner of the building, without specifying the floor it's on. The address 1100XX carries 4 bits of information and denotes any room in the northeast apartment on the fourth floor, without specifying the room. Notice that the more specific the address, the more the information. In like manner, the more specific a gene or a protein, the more information it contains.

A protein whose performance would be affected by a change in *any one* of its amino acids is *very* specific. If its performance would not be affected by a change in *some* of them, the protein would be *less* specific. Often, an enzyme is very specific: a change of any one of its amino acids results in some sort of change in the enzyme's performance.

I can illustrate the relation between specificity and information in enzymes by drawing an analogy from substrate molecules to English words. Suppose, for example, we think of what we might call a *word-enzyme* that matches to all words containing the letter string "ghtsha". There is only one word in the English language that has that sequence of letters — the word *nightshade*. If the string "ghtsha" were used for a match it would be highly specific.

If we reduce the information in the match string by dropping the "a" at the end, and match only to "ghtsh," the match becomes less specific since there are now two more words that match, namely, *nightshirt* and *lightship*. These matches, however, would be "weaker" than the previous one because they match to only five letters, whereas the previous match was to six.

If we further reduce the information in the match string by dropping the last "h" to get "ghts", the match would lose a little more strength because it uses only four letters. But it would lose a lot more in specificity because there would now be about a hundred matching words, particularly nouns in the plural form, such as *lights*, *weights*, and *thoughts*. Thus we see that reducing the specificity of the matching sequence of an enzyme goes hand in hand with reducing the information in it.

The most noticeable type of change in an enzyme is in its activity as a catalyst. But there could be other kinds of changes too. For example, there might be a change in how well the enzyme attaches to its substrate, or in the way its control mechanism works. A mutation almost anywhere in an enzyme that's controlled through an allosteric* site will affect some aspect of its performance. Studies of enzymes *in vitro*† cannot show up all the effects of a change in an amino acid. Only through careful studies *in vivo*‡, where the enzyme can express its full capability, can one hope to reveal all the effects of changing one of its amino acids.

———

All point mutations that have been studied on the molecular level turn out to reduce the genetic information and not to increase it. Let's examine what's known about the resistance of bacteria to antibiotics and of insects to pesticides.

Some bacteria have built into them at the outset a resistance to some antibiotics. The resistance comes from an enzyme that alters the drug to make it inactive. This type of resistance does not build up through mutation. J. Davies and his colleagues at the University of Wisconsin expressed the opinion that the purpose of the enzyme may not be to offer resistance to the drug. They did not profess to know its primary purpose, but they considered it to be directed toward attacking small molecules involved in some other, but so far unknown, cell function [Davies et al. 1971]. In their opinion, drug resistance in these cases may be only fortuitous. On the other

———

* This term is defined in Appendix D.
† *In vitro* is Latin for "in glass," used to designate studies made in a test tube or other laboratory environment, as distinguished from studies made within the living organism.
‡ *In vivo* is Latin for "in life," used to designate studies made within the living organism.

hand, our commercial antibiotics are the natural products of certain fungi and bacteria [Aharonowitz and Cohen 1981]. One might therefore expect that some bacteria would be endowed with an enzyme providing resistance to them.

Bacteria that are not resistant can become resistant through infection by a virus that carries the gene for resistance. The virus may have picked up the gene from a naturally resistant bacterium. Also, bacteria can be deliberately made resistant by artificially introducing into their DNA the gene encoding the enzyme. Scientists today can transfer sections of DNA from one organism to another. The gaining of antibiotic resistance in this way is not an example of how evolution might add information. The genome of the bacterium that acquired the resistance does indeed gain information. But there is no gain for life as a whole. The resistant gene already existed in some other bacterium or virus.

But some bacteria can mutate to become resistant to a drug to which it had been sensitive. In these cases the function is new. Could such a mutation demonstrate neo-Darwinian evolution?

Scientists have studied how streptomycin and other mycin drugs keep bacteria from growing, and how a point mutation makes bacteria resistant to the drug [Davies et al. 1971, Davies and Nomura 1972]. They found that a molecule of the drug attaches to a matching site on a ribosome of the bacterium and interferes with its making of protein, as shown in Fig. 5.3. With the drug molecule attached, the ribosome is unable to put the right amino acids together when it makes protein. It makes the wrong proteins. It makes proteins that don't work. The bacterium then can't grow, can't divide, and can't propagate.

The ribosomes of mammals don't have the site at which the mycin drugs can attach, so the drugs can't harm them. Be-

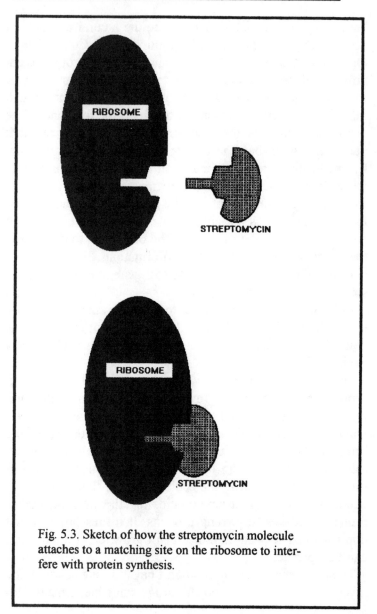

Fig. 5.3. Sketch of how the streptomycin molecule attaches to a matching site on the ribosome to interfere with protein synthesis.

cause the mycins can stop bacterial growth without harming the host, they make useful antibiotics.

A point mutation makes the bacterium resistant to streptomycin by *losing* information. If a mutation in the bacterium should happen to change the ribosome site where the streptomycin attaches, the drug will no longer have a place to which it can attach. Fig. 5.4 shows schematically how a change in the matching site on the ribosome can prevent a streptomycin molecule from fitting onto the ribosome and interfering with its operation. The drug molecule cannot attach to the ribosome, so it cannot interfere with its making of protein, and the bacterium becomes resistant.

You can see from Fig. 5.4 that the change could be in several different places on the matching site and still grant resistance to the bacterium. Any one of several changes in the attachment site on the ribosomal protein is enough to spoil its match with the mycin. That means that a change in any one of several DNA nucleotides in the corresponding gene can confer resistance on the bacterium. Several different mutations in bacteria have indeed been found to result in streptomycin resistance [Gartner & Orias 1966]. We see then that the mutation reduces the specificity of the ribosome protein, and that means *losing* genetic information. This loss of information leads to a *loss of sensitivity* to the drug and hence to resistance. Since the information loss is in the gene, the effect is heritable, and a whole strain of resistant bacteria can arise from the mutation.

Although such a mutation can have selective value, it decreases rather than increases the genetic information. It therefore cannot be typical of mutations that are supposed to help form small steps that make up macroevolution. Those steps must, on the average, add information. Even though resistance is gained, it's gained not by adding something, but by losing it. Rather than say that the bacterium gained resistance to the antibiotic, we would be more correct to say that it lost its sensitivity to it. It lost information.

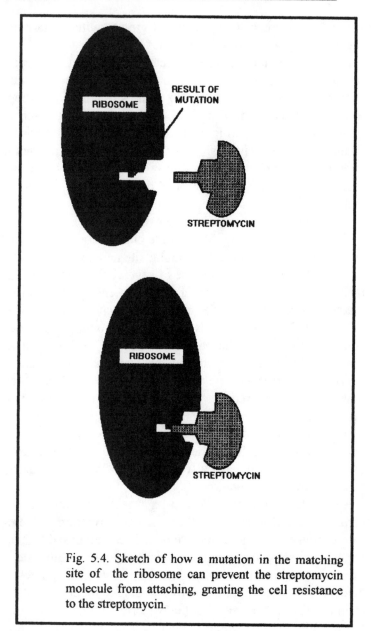

Fig. 5.4. Sketch of how a mutation in the matching site of the ribosome can prevent the streptomycin molecule from attaching, granting the cell resistance to the streptomycin.

The NDT is supposed to explain how the information of life has been built up by evolution. The essential biological difference between a human and a bacterium is in the information they contain. All other biological differences follow from that. The human genome has much more information than does the bacterial genome. Information cannot be built up by mutations that lose it. A business can't make money by losing it a little at a time.

———

Scientists have found how a mutation can make an insect resistant to an insecticide. A molecule of the insecticide DDT acts by binding itself to a specific matching site on the membrane of the insect's nerve cells. In this position it prevents the nerve from functioning properly. When enough of the insect nerve cells have DDT molecules bound to them, the nervous system breaks down and the insect dies [Beeman 1982].

How does an insect become resistant? It becomes resistant by losing its sensitivity to the DDT. This loss is the result of a mutation that changes the site on the nerve cell at which the DDT molecule binds, preventing the DDT from binding [Tanaka et al. 1984]. Any mutation that spoils the match between the DDT and the nerve cell will make the insect resistant. As with bacteria, resistance can come by reducing the specificity of the protein of the nerve cell.

———

Changing an amino acid in a protein very often affects the way the protein functions.* An organism is generally well adapted to its niche. Its proteins are well suited to carrying out their functions. A change in one of its proteins is then likely to degrade the organism in some way. In particular, when an organism becomes resistant to a drug through a

---

* George Wald (Nobel Laureate 1967) has said, "One is hard put to find a single instance in which a change in one amino acid [in a protein] does not change markedly its properties." [Wald 1967, p. 19]

143

change in one of its proteins, it is likely to become less fit in some other way. Of course, so long as the drug is present, the organism has to be resistant to survive, even at the price of being less fit in another way. But when the drug is removed, the nonresistant type is again more adaptive.

A mutation in bacteria that makes it resistant to streptomycin reduces the specificity of a protein in the ribosome. When the ribosome becomes less specific, its performance is degraded. T. K. Gartner and E. Orias from the University of California in Santa Barbara reported some time ago that the mutations that make bacteria resistant to streptomycin degrade their ribosomes [Gartner and Orias 1966]. The mutation makes the ribosome slower than normal in translating some of the RNA codons into protein.

Degrading side effects have also been noted in insects that have become resistant to insecticides. M. W. Rowland from the Rothamsted Experimental Station in Hertfordshire, England has reported that mosquitoes that have become resistant to dieldrin are less active and slower to respond to stimuli than are other insects [Rowland 1987]. Their resistance to the insecticide is thus bought at the price of a more sluggish nervous system. The information loss on the molecular level then appears as a loss in the performance of the insect.

Geneticists distinguish between traits they call *qualitative* and those they call *quantitative*. Whereas a change in a qualitative trait shows up as an abrupt change in the phenotype, changes in a quantitative trait often seem to be continuous. For example, the change of a normal corn plant to a dwarf in one step is a qualitative change. A small change in the height of normal corn is a quantitative change. Man has for a long time been manipulating quantitative traits in living organisms, breeding and selecting plants and animals to increase his economic gain. He has bred wheat to yield more and better food value, fruits and vegetables to make them larger, and cows to give more milk or yield more beef.

Quantitative traits such as these build up little by little over many breeding generations. Sometimes a small change in a trait is caused by the environment rather than by a genetic change, and it is therefore not heritable. In other cases the change is a true genetic change, and is heritable. Breeders try to exploit the heritable changes in plants and animals to build up the traits they want.

In many breeding experiments, the variations that appear don't come directly from mutations. Often the variation was already in the population. Sometimes a change could result from a recombination of genes as described in Chapter 2. But the ultimate source of the variation in the population, according to the NDT, must have been mutations that occurred in the past.

In many, if not all, experiments with the selection of quantitative traits, mutations at several different sites on one or more genes contribute in a cumulative way to an increase or decrease in some parameter [Herskowitz 1962]. Mutations can build one on the other. Vegetables can increase in size with each successive mutation. Moreover, the variation appears to be limitless in some breeding experiments that have been in progress for several decades; and the trend in the trait still continues.

Could mutations such as these serve as prototypes of Darwinian macroevolution? To answer this question we have to see how these mutations act on the organism. What is known about these mutations and how do they affect the phenotype?

As noted in Chapter 2, regulatory proteins control the making of protein in the cell. The regulatory proteins exert either a positive or a negative control. When all of them act in concert, they produce a tight and finely-tuned control. Under normal conditions, the rate at which protein is made is kept at a level that's best for the organism's well being. The positive control factors tend to raise protein production, and the negative factors tend to lower it. The tension between them makes the control sensitive to change, and permits it to act rapidly and accurately.

Mutations that increase the food-protein yield in wheat have been discovered in genes that control the making of the protein. They have been found to increase protein production by degrading a negative control factor [Konzak 1977]. One can reasonably suppose that other quantitative traits such as the size of vegetables and the amount of milk that cows give are also affected by mutations in this way.

One such mutation after another can progressively increase the rate at which the cell makes food protein. But they do it at the cost of upsetting the cell's fine control of protein production. A change in any one of many nucleotides in a regulatory gene can degrade the repressor protein encoded by the gene. Each change that further degrades the repressor protein will further increase the protein yield. The changes lose genetic information as they reduce the specificity of the regulatory protein.

Any one of many possible mutations can degrade the regulatory gene and weaken the bond between the repressor and the operator on the gene to be controlled. A second mutation can weaken the bond even more. In this way, one mutation after another can build up to make the repressor less and less effective. The cell will then make more and more protein.

If there are several genes encoding food protein in the cells of a grain of wheat, there can be many sites at which mutations could increase protein yield. Geneticists have long known that quantitative traits are often governed by more than one gene [Herskowitz 1962]. This seems to be the rule rather than the exception. There can be many combinations of mutations among the various genes, and they can lead to many possible levels of protein yield. The absolute limit of protein yield would obtain when all repressor functions have been fully disabled, if the plant can survive in that condition.

For example, the internode length of the spike in the barley plant has been shown to be affected by mutations in at least 26 different sites [Persson and Hagberg 1969]. Some of these sites have more than 30 alleles. If each of the 26 sites had on the average only 15 alleles then there would be a total of 15^{26}, or 3.8×10^{30} different possible combinations of mutations.*

With 3.8×10^{30} different combinations of internode length, recombination with selection could go on using new combinations for a long time. Of course, many of the combinations may well yield the same apparent result in the internode spike length. But, still, there are likely to be very many different lengths. They might even appear to be inexhaustible over the duration of an experiment. There would of course be a limiting value for the internode spike length. That would occur when the control is all in one direction.

There are several kinds of controls that together govern the size of an organism. The sizes of plants and animals are normally influenced by hormones that stimulate growth and by other hormones that repress it. The amount of these hormones in the system is determined by the system controls. You can readily see how a degrading mutation can make a plant or animal become larger. The organism will grow larger if a mutation degrades the hormone that represses growth. It will also grow larger if a mutation degrades a molecule that represses the making of the hormone that stimulates growth. One mutation after another can progressively degrade a repressor protein. As they do so, they would produce a larger and larger organism. A degrading mutation can also make a plant or animal grow to smaller size. A mutation that degrades an activator of a gene for a hormone that stimulates

---

* Lest one think that vast amounts of variation can be stored in this manner, one must note that even 3.8×10^{30} is only an insignificant fraction of the number of different possible amino-acid sequences of 26 ordinary-sized protein molecules. If each of the 26 proteins had 300 amino acids, then there would be $20^{7,800}/26!$ different possible arrangements of amino acids, which is about $10^{10,121}$.

147

growth would stunt the growth. So would a mutation that degrades a repressor of a gene for a hormone that represses growth.

Mutations in control genes tend to have harmful side effects. They usually disturb the normal balance in the organism's functioning. In humans, for example, too much growth hormone from the pituitary gland can lead to gigantism and to diabetes [Landau 1967]. In grain, selection for high protein content has been found to result in less starch per seed and less grain per planting [Brock 1980]. Dairy cattle bred for high milk production turn out to be less fertile than normal cattle [Hermas et al. 1987]. Nevertheless, cattle raisers and farmers are willing to accept these side effects to get a more productive animal or a larger plant. Edward O. Wilson of Harvard University has said,

> Artificial selection has always been a tradeoff between the genetic creation of traits desired by human beings and an unintended but inevitable genetic weakness in the face of natural enemies [Wilson 1992].

The negative side effects appear because information has been lost from the gene — because a regulatory gene has become less specific.

None of the kinds of mutations mentioned above can serve as a prototype for the mutations needed for neo-Darwinian evolution. Although there are circumstances where point mutations are good for the organism, all known point mutations lose information. Some microevolution may indeed occur this way. But a mutation that loses information, even if it's good, cannot be a typical member of a chain of mutations for cumulative selection. The prototype of the mutations that are supposed to make up neo-Darwinian macroevolution must be one that adds a small amount of information.

———

There have been some experiments whose results seem to defy the laws of information theory and show the impossible. We have seen (in Chapter 3) that on the average a single step of evolution can't add much more than one bit of information to a genome. Yet some experiments have shown steps that seem to add a lot of information.

Experiments with soil bacteria have shown a phenomenon that some geneticists have pointed to as an outstanding example of the basic processes of evolution. Bacteria grown in culture have shown they can learn to live and grow on new substances that they originally could not use [Mortlock 1982]. The bacteria were put under strong selection pressure. The experimenters denied them their normal nutrients, which are either ribitol or D-arabitol, and tried to get them to grow on an unnatural substitute. They tried several unnatural sugars, and met with success when they used the sugar xylitol. Xylitol is a sugar very similar to ribitol and D-arabitol, but does not occur in nature. Mutants appeared that could live on xylitol, and cultures were grown from them. Several experiments of this kind have been reported [Lerner et al. 1964, Wu et al. 1968, Rigby et al. 1974, Burleigh et al. 1974, Inderlied and Mortlock 1977, Thompson and Krawiec 1983].

The experiments, moreover, showed that not only one, but a series of mutations could occur that would first establish a new enzyme to metabolize xylitol, and then progressively improve it. Isolation and culturing of soil bacteria led to a series of three mutations. The experimenters started with the wild-type bacteria, which could not grow at all on only xylitol. A mutant soon appeared in the culture that could grow on xylitol. They cultured this mutant and labeled the strain X1. They found that X1 grew on xylitol at about one ninth the rate of the wild type on ribitol.

In culturing the X1 strain on xylitol, they found a second mutant appeared that could grow 2.5 times as fast on xylitol as did X1. They cultured this second mutant and labeled the strain X2.

In culturing X2 on xylitol, they found a third mutant that grew almost twice as fast as did X2. They cultured it and labeled the strain X3. The three mutations were studied and all were found to be single nucleotide substitutions.

These experiments show that bacteria can sometimes find other ways of getting what they need when their normal nutrients are denied them. Moreover, they did it through a series of point mutations. The first mutation enabled the bacteria to metabolize a nutrient they could not use before. Two more mutations improved on the ability.

These experiments look like neo-Darwinian evolution in action. The experiments seem to show bacteria evolving through a series of three small steps. Can this short series of steps be part of a chain of cumulative selection? Can these three steps, performed in only a few months under artificial selection, serve as a model for a much longer series of steps under natural selection that might take a million years and lead to macroevolution? Could these steps show the sort of evolution that primitive bacteria underwent in developing their enzymes?

If we look into these experiments in detail, we see that no new information got into the genome. Indeed, it turns out that each of those mutations actually *lost* information. They made the gene less specific. Therefore, none of them can play the role of the small steps that are supposed to lead to macroevolution.

Normally the wild type feeds on ribitol. The cell takes in the ribitol from the outside, and breaks it down in a series of steps, with a special enzyme for each step. The first of these enzymes is ribitol dehydrogenase (RDH).

Ribitol is a sugar residue found in the soil. Xylitol, on the other hand, is not found in nature, but its structure is similar to ribitol. Xylitol and ribitol are made up of the same atoms in almost the same arrangement. The difference between them is slight, yet the cell's RDH enzyme is specific to ribitol. Fig. 5.5 shows the structures of ribitol, xylitol, and another nonnatural sugar residue, L-arabitol (which I shall soon discuss). The figure does not show the arrangement of the

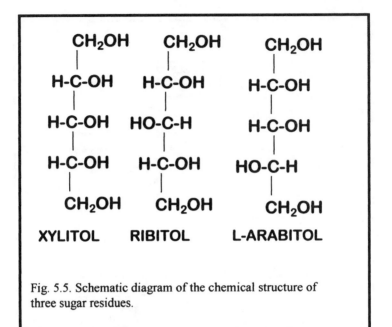

$$
\begin{array}{ccc}
\text{CH}_2\text{OH} & \text{CH}_2\text{OH} & \text{CH}_2\text{OH} \\
| & | & | \\
\text{H-C-OH} & \text{H-C-OH} & \text{H-C-OH} \\
| & | & | \\
\text{H-C-OH} & \text{HO-C-H} & \text{H-C-OH} \\
| & | & | \\
\text{H-C-OH} & \text{H-C-OH} & \text{HO-C-H} \\
| & | & | \\
\text{CH}_2\text{OH} & \text{CH}_2\text{OH} & \text{CH}_2\text{OH} \\
\textbf{XYLITOL} & \textbf{RIBITOL} & \textbf{L-ARABITOL}
\end{array}
$$

Fig. 5.5. Schematic diagram of the chemical structure of three sugar residues.

atoms in three dimensions, but it gives you an idea of how similar these sugars are to each other. Because ribitol and xylitol are so much alike, the same RDH enzyme that works on ribitol works to a small extent also on xylitol. It breaks down xylitol to make the same product it makes from ribitol. After this one step, all other steps in the metabolism of ribitol and xylitol are the same. But because the RDH is highly specific to ribitol, it works only poorly on xylitol.

Some of the reasons why the normal cell cannot use xylitol have to do with the cell's genetic control system. The normal way this control system works is shown in Fig. 5.6. The enzyme RDH is made when gene Y is turned ON. Gene Y is normally OFF, but is turned ON by the presence of ribitol. Moreover, molecules, including ribitol, cannot easily get into the cell through the cell wall. They only get into the cell if the cell permits them entry. If the cell wants a molecule taken inside, it uses a special transport mechanism, known as a

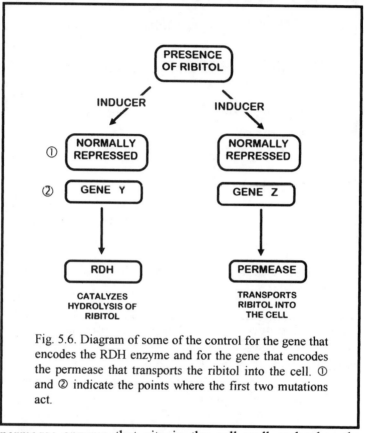

Fig. 5.6. Diagram of some of the control for the gene that encodes the RDH enzyme and for the gene that encodes the permease that transports the ribitol into the cell. ① and ② indicate the points where the first two mutations act.

permease enzyme that sits in the cell wall and takes the molecule in.

There are three problems that prevent the normal cell from using xylitol. They are:

1. The cell does not make RDH, until ribitol induces the RDH gene to transcribe its code from the DNA to mRNA. Xylitol will not induce it. Therefore there will be no RDH when there is no ribitol.

2. Although RDH has some activity on xylitol, that activity is much lower than it is on ribitol.

3. Ribitol and xylitol, as well as other molecules, cannot get into the cell by themselves. They do not diffuse well through the cell wall. Ribitol is normally carried into the cell by a special transport system that is specific to ribitol. That system is mainly a permease enzyme which sits in the cell wall. The cell does not make the permease enzyme, however, until ribitol induces its transcription. There is no transport system for xylitol.

The X1 mutant overcame the above problems to some extent through a point mutation shown at position ① in Fig. 5.6. This mutation is in the gene that regulates the making of RDH. This regulatory gene encodes a protein whose job it is to repress the making of the RDH. The mutation did not change the enzyme molecule itself. What it did was destroy the repressor protein. As a result, there was no repressive control in the making of RDH. The gene could then transcribe RNA without having to be induced, and it did so at the maximum rate. RDH was then made in such abundance that, in spite of its low activity on xylitol, the cell could function. The cell could function because:

1. Since the mutation blocked the repression of the gene transcription, the cell made the RDH without it having to be induced.

2. Unrepressed, the gene made the RDH enzyme at its maximum rate. The large amount of RDH that was made helped make up for its low activity on xylitol.

3. Although the transport system does not bring xylitol into the cell, a small amount does get in by diffusion.

The X1 did not have a perfect solution to the three problems. Indeed it contributed nothing at all to the third problem. It therefore grew much more slowly on xylitol than the wild type grows on ribitol. Nevertheless, X1 can grow on xylitol alone, and the wild type cannot. But the benefit of the mutation came through a *loss* of information.

The X2 strain resulted from a point mutation in an X1 bacterium. This second mutation is at ② in Fig. 5.6. It changed the enzyme itself and raised its activity on xylitol. Because of the higher activity, the growth rate of X2 on xylitol was about 2.5 times that of X1.

Because the mutation made the enzyme more active on xylitol, you might think it made the enzyme more specific. You might therefore think that in this case genetic information increased.

But it turns out that the mutation that led to the X2 strain is just another example of a mutation making the enzyme *less* specific. Brian Hartley and his group at the Imperial College in London made a study of this enzyme [Burleigh, et al. 1974]. They compared its activity in the X2 mutant with that of the enzyme of the wild-type. They measured the activity of the enzyme on ribitol, xylitol, and L-arabitol, another unnatural substrate.

They found that, compared to the wild type, the mutant enzyme was

- less active on ribitol,
- more active on xylitol, and
- more active on L-arabitol.

Fig. 5.7 shows a comparison of the reaction rates of the two forms of the enzyme for the three substrates. The plot was made from the numerical data in the Burleigh paper. The enzyme in the X1 strain is denoted in the figure by A. The enzyme in the X2 strain is denoted by B. The vertical scale shows the reaction rate, or the rate of catalysis. Note that the mutation that transformed the X1 strain into X2 broadened the range of substrates that the enzyme could catalyze. The enzyme of the wild type (A) has a reaction-rate curve higher and narrower than that of the mutated type (B). A very specific enzyme would show a high narrow plot of reaction rate. A less specific one would show a low wide plot. Fig. 5.7 shows that the reaction rate of B is less specific than that of

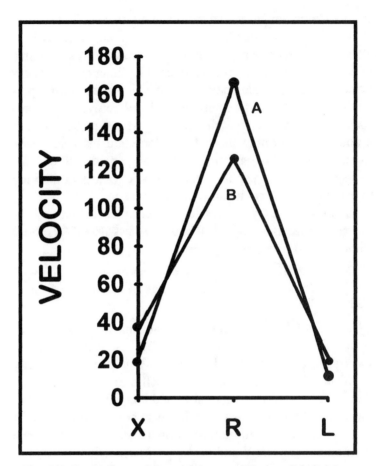

Fig. 5.7. Comparison of the activity to wild-type ribitol dehy-drogenase with the mutant form on three different substrates. Velocity scale is in units/mg. X = xilitol, R = ribitol, L = L-arabitol.

A. The new enzyme could accept a wider range of molecules as substrates than the old one. Thus the mutation made the enzyme *less* specific, not more, and reduced the information in the genome.

Roger Bone and his colleagues at the University of Cali-fornia have shown a clear example in which a change in one amino acid in a natural enzyme made it less specific [Bone et al. 1989]. They changed one amino acid in an enzyme called alphalytic protease, and found that the enzyme was then able to act on a wider class of substrates. The function of this enzyme is to cut peptide chains at a unique place within a particular sequence of four amino acids. Wherever that foursome occurs in a peptide chain, the enzyme cuts the chain at a unique point within those four. Bone and his team made two changes in the enzyme. Each change made the enzyme less specific. It cut peptide chains at other places in addition to just the one where the original enzyme cut.

The changed enzyme was less specific, so it had less information than the original. It was also less useful. As those authors noted [Bone et al. 1989]:

> One of the fundamental functions of an enzyme is to provide specificity by limiting the range of substrates which are catalytically productive.

One might have thought that if an enzyme's activity on some substrate increases, it must be because it became more specific to that substrate. But we see that is not necessarily so. If an enzyme becomes more specific to a substrate U, it should become more active on U and less active on the other substrates V and W. Also the other way around; if it becomes less specific on U, it should become more active on some other substrates, say V and W. Fig. 5.7 shows this.

The specificity of an enzyme to its substrate is no less important to the cell than the level of its activity. An enzyme that will accept any molecule as its substrate can be harmful. For an enzyme to be useful to the cell, it must limit its activity to its proper substrate.

Some of the specificity of an enzyme lies in the kind of reaction it catalyzes, and some lies in the kind of substrate it will act on. The active site of an enzyme can be considered to consist of two parts. One part recognizes and binds to the substrate, and the other catalyzes the reaction on it [Darnell et al. 1986, p. 65].

The X2 mutation evidently degraded the place on the enzyme to which the substrate binds. It weakened the bond between the enzyme and ribitol. The mutation also made the enzyme less able to discriminate against other substrates such as xylitol and L-arabitol. It made the enzyme more active on xylitol than it had been, but not because it added information. The mutation did not make the enzyme more specific for xylitol. It made it less specific. The enzyme's increased activity on xylitol, as well as on other substrates, indicates its loss of specificity.

S. A. Lerner and his colleagues at Harvard Medical School have shown that the mutated enzyme of X2 is less stable than that of the wild type [Lerner et al. 1964]. Typically, when an enzyme loses information its function is degraded. The mutation leading to the X2 strain is a point mutation and is indeed an example of a small genetic change. It is an example of microevolution, but it can't be typical of a step in macroevolution, because it adds no information to the genome.

The X3 strain came from X2 through a point mutation shown at ③ in Fig. 5.8. This mutation disabled the gene that regulates the transcription of the transport enzyme that carries a nutrient into the cell. In this case the nutrient is D-arabitol. The transport system turns out also to be able to bring xylitol into the cell. But normally the transport enzyme isn't made unless its gene is turned ON by the presence of D-arabitol. In the absence of D-arabitol a repressor protein keeps the gene OFF. So even though the D-arabitol transport enzyme can also work on xylitol, it doesn't normally get made unless it gets induced by D-arabitol. The mutation leading to X3 disabled the gene encoding the repressor protein, so that it could not repress the gene for the transport enzyme. As a result the transport enzyme was made at the maximum rate and in large amounts without regulation. There was then no need for its induction by D-arabitol. Xylitol then gets a free ride into the cell on the transport enzyme intended for D-arabitol.

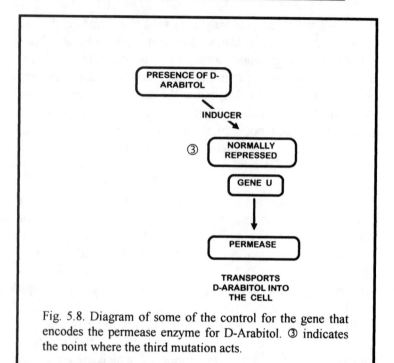

Fig. 5.8. Diagram of some of the control for the gene that encodes the permease enzyme for D-Arabitol. ③ indicates the point where the third mutation acts.

Much more xylitol could then enter the X3 cell than could enter the X1 or X2 cell. Therefore X3 could feed on xylitol better than X2 could. As with the X1 and X2 mutations before it, the mutation leading to X3 also reduced the specificity of a gene and lost information. Similar results have been shown with other bacteria and with other nutrients.

There are, however, several examples of mutations that permit bacteria to live on a new nutrient and seem to add a lot of information. Some experiments show the introduction of an entirely new enzyme. One group of scientists started with bacteria that did not have the beta-galactosidase enzyme needed to break down lactose [Campbell et al. 1973, Hartl and Hall 1974, Hall 1978, Hall 1982]. They cultured the bacteria on lactose. The bacteria could not grow and multiply unless they learned to use lactose. In the culture, a mutant appeared that could metabolize lactose. This mutant multiplied while the others did not. The descendants of this mutant were extracted from the culture and used to start another culture. This new culture could live on lactose even though the bacteria did not have the beta-galactosidase enzyme. Several other examples of this kind of evolution have been reported and were described in review articles by Hall [1982] and by Mortlock [1982].

Some recent experiments have shown that bacteria can mutate to produce new functions when they are needed. The mutations do not occur when they are not needed [Hall 1982, Cairns et al. 1988, Hall 1988]. The mutations were not point mutations. The results seem to contradict the neo-Darwinian theory, and as a result they led to an immediate flurry of comments in the literature.* Hall's and Cairn's results pose a challenge to neo-Darwinian theory. These phenomena themselves call for a new theory. I shall take up this subject in Chapter 7.

———

I have indicated in this chapter that there is no evidence that genetic information can build up through a series of small steps of microevolution. Mutations needed for these small steps have never been observed. By far, most observed mutations have been harmful to the organism. We have seen that there are some point mutations that, under the right circumstances, do give the organism an advantage. There are point mutations that make bacteria resistant to antibiotics. There are some that make insects resistant to insecticides. There are some that increase quantitative traits in farm plants and animals. But all these mutations reduce the information in the gene by making a protein less specific. They add no information, and they add no new molecular capability. In-

---

* Some of these comments can be found in Stahl [1988], Benson [1988], Partridge and Morgan [1988], Charlesworth et al. [1988], Graffen [1988], Holliday and Rosenberger [1988], Van Valen [1988], Danchin [1988], Tessman [1988], Lenski et al. [1989], Lenski & Mittler [1993], and Davis [1989].

deed, all mutations studied destroy information. None of them can serve as an example of a mutation that can lead to the large changes of macroevolution.

The neo-Darwinians would like us to believe that large evolutionary changes can result from a series of small events if there are enough of them. But if these events all lose information they can't be the steps in the kind of evolution the NDT is supposed to explain, no matter how many mutations there are. Whoever thinks macroevolution can be made by mutations that lose information is like the merchant who lost a little money on every sale but thought he could make it up on volume.

With all the experimental results in genetics, there is no evidence for the Darwinian Assumption 1 of Chapter 4. The neo-Darwinians presume that a long chain of random changes can lead to a large evolutionary change. This conjecture is an essential point of their theory. The lack of evidence for this conjecture means we have to reject Assumption 1 of Chapter 4. We therefore have to reject the entire neo-Darwinian theory of evolution.

Not even one mutation has been observed that adds a little information to the genome. That surely shows that there are not the millions upon millions of potential mutations the theory demands. There may well not be any. The failure to observe even one mutation that adds information is more than just a failure to find support for the theory. It is evidence *against* the theory. We have here a serious challenge to neo-Darwinian theory.

Chapter 6

THE WATCHMAKER'S BLINDNESS

ABOUT ten years ago a book was published called *The Blind Watchmaker* [Dawkins 1986]. The author is Richard Dawkins, a zoologist at Oxford University, who believes with a passion that life evolved all by itself. In his book he did his best to instill that same passionate belief into his readers. The book is a missionary treatise in which the author hoped to convert the nonbelievers. His goal seemed to be to convince his readers that life is the product of chance events, that it is the natural consequence of physical laws. He argued that the neo-Darwinian theory explains well how life arose and developed, and there's no need to invoke a Creator for that purpose. He tried to convince the reader that the universe was not created by some Supreme Being, but it came into being in a natural way according to the natural laws by which the universe runs.

Recall that the Argument from Design (AFD) draws an analogy from an artifact, like a watch, to living organisms. The AFD starts by inferring the existence of a watchmaker from the existence of a watch. In Paley's [1817] words:

> Contrivance must have had a contriver, — design, a designer ... [p. 11]

The Argument then proceeds to infer the existence of a Creator from the existence of life,

> ... the contrivances of nature surpass the contrivances of art, in the complexity, subtlety, and curiosity of the mechanism. [p. 13].

The Paley argument is *a fortiori*. As it is practically impossible for an inanimate object to come into being by chance, how much more so for life. But Dawkins pointed out that living things, which can reproduce, can develop gradually through natural selection from something simpler. They can *evolve* by cumulative selection. The chance that they could gain complexity this way would be much greater than the chance that something complex could emerge from simple components coming together all at once. The chance that a frog could evolve from a single cell would then be greater than the chance that a watch would assemble itself from its raw materials.

Dawkins concluded that the neo-Darwinian postulate of random variation, followed by natural selection, refutes the AFD. Life is the result of chance rather than design. Dawkins's analogy to the watchmaker is not the Creator, as Paley had it. It is blind chance. Or as Dawkins put it, "the watchmaker is blind."

———

Dawkins is what I have called a brady. He believes that macroevolution takes place gradually through long chains of many small steps called *cumulative selection*. Each step in a chain is a small adaptive change followed by natural selection. Dawkins's point here is that there is a much larger chance for cumulative selection to occur than for one large adaptive step. That's true. Nevertheless, as I have shown, the chance of cumulative selection occurring is too low to serve as the basis of a theory of evolution.

Dawkins talked about chance, but he didn't calculate the chance of anything. Nor did he cite anyone who did. He just assumed that cumulative selection could lead to macroevolution. He *assumed* what I have shown to be impossible. He said, without justification:

> Each successive change in the gradual evolutionary process was simple enough, relative to its predecessor, to have arisen by chance. [Dawkins 1986, p. 43]

He granted that a chance of 10^{-190} for getting the hemoglobin molecule in one step is too small for one to expect it to occur. But he didn't suggest what *would* be acceptable. He did no more than state that the chance of getting hemoglobin by cumulative selection is greater than getting it in one step.

In Chapter 4, I discussed the evolution of the horse, which is typical of evolution according to the bradys. The chance of occurrence of one of the small steps that are supposed to build up to a new species is too small for evolution to work, unless two stringent assumptions are satisfied. We have seen several examples for which the assumptions are not satisfied. There are no examples for which the assumptions *are* satisfied. We must conclude that Dawkins was mistaken in what he wrote about the power of cumulative selection.

Dawkins made a major error that invalidates his conclusions. He assumed the steps that make up the process of cumulative selection could be as *small* as you might wish. He implied by this that the steps would be as *likely to happen* as you might wish. This tacit assumption is wrong. There are stringent requirements on the steps: they must *each have a selective advantage*, and they must each on the average *add a small amount of information* to the genome. The requirement of selective advantage prevents the steps from being as small as you like. Dawkins spent a chapter to show "the power of this cumulative selection." He gave no space at all to the question of whether or not such cumulative selection is *possible*. To answer such a question, one has to compute.

Mutations cannot be as small as you'd like. Their minimum size is a point mutation of one DNA nucleotide. My analogy of trying to improve a novel by evolution showed that a single change usually requires other changes to go with it. And because each step must satisfy some minimum requirements, one cannot assume that the steps can be arbitrarily small.

There is still a lot we don't know about living organisms. We know little about how they interact with each other. We know little about the selection pressures on populations, and about how they respond to those pressures. But we must not use our ignorance as a license to create artificial facts to support a favorite theory.

———

Dawkins set for himself the task of showing that all the complexity of life is the result of natural laws. He therefore also dealt with the origin of life. Dawkins noted that cumulative selection can work only on systems that can reproduce. He defined life as a self-replicating system. How did he suppose the first such system could have arisen? Clearly, it could not have arisen through cumulative selection.

Dawkins recognized the problem, and in his Chapter 6 he addressed the question of how a self-replicating system could arise. It had to arise all at once. We have seen that the chance of such an event is very small, and herein lies the problem of the spontaneous origin of life.

He did not, however, make a convincing case for the origin of life. All suggestions so far made for the origin of life are based on random events. To test them one must at least find the chance that the required events could happen in the time available. Dawkins mentioned two mechanisms: the theory of the "primeval soup" and the Cairns-Smith theory [Cairns-Smith 1984, 1985]. He discussed the latter in some detail. Since no one has computed, for either theory, the chances of the events occurring, Dawkins could not tell us what those chances are. The mechanisms of both theories, however, have every appearance of being very improbable — even to the point of being *impossible*.

The origin of life is a harder problem than that of life developing once it got started. That's because one can't make use of natural selection for the beginning of life. Therefore, one should strive for a theory that has the present complexity of life develop from as simple an origin as possible.

Dawkins asked that we drop our intuitive feeling for probability, or chance. Our intuition, he said, is not suited to dealing with something as improbable as the origin of life. He noted that the origin of life may be highly improbable at any one place at any one time. Yet, it had a lot of time and space going for it, and that would be enough to make it more likely than not. He, therefore, advised us to try to change the way we look at probabilities. We must stop thinking of the relatively large probabilities with which we are used to dealing. We should, instead, try to think of the very small probabilities that might suit a hypothetical alien being who lives millions of years.

Dawkins's advice shows that he didn't understand probability. For example, drawing a red ball from an urn containing both red and black balls is an event. A coin flipped eleven times and turning up nine heads and two tails is an event. Life occurring spontaneously at least once in 10^{20} planets in a billion years is also an event. If we can compute the chance of this event, we can comprehend it in just the same way we comprehend the chance of drawing a red ball from an urn. If a theory predicts that chance to be one in a million or less, then the theory *does not account for* the origin of life. Actually, those theories from which we can compute a number, yield a chance much *less* than one in a million.

Dawkins's long-lived alien doesn't make any sense and doesn't help make anything clear. Dawkins slipped into a common error with his long-lived alien. Dawkins said that a creature that lives millions of years would have a different feeling for the meaning of the chance of an event than we have. If the alien lives a hundred million years, he could have played very many hands of bridge. Then, Dawkins said, it would not be unusual for him to see a "perfect" bridge hand where each player was dealt thirteen cards of the same suit.

> They will expect to be dealt a perfect bridge hand from time to time, and will scarcely trouble to write home about it when it happens.

He's wrong. One can easily calculate the chance of Dawkins's alien experiencing a perfect bridge hand at least once in his lifetime. The chance of getting such a hand in one deal is 4.47×10^{-28}. If the alien plays 100 bridge hands every day of his life for 100 million years, he would play about 3.65×10^{12} hands. The chance of his seeing a perfect hand *at least once in his life* is then 1.63×10^{-15}, or about one chance in a quadrillion. That's less than Dawkins's chance of coming to New York for two weeks and winning the lottery twice in a row. Would he bother to write home about it?

Dawkins's error is one that evolutionists often make. Many of them have fallen into that trap. They think the earth's age is long enough for anything to have happened. When one deals with events having small probabilities and many trials, one should multiply the two numbers to determine the probability[*]. One should not just stand gaping at the long time available for trials, ignore the small probability, and conclude that anything can happen in such a long time. One has to calculate.

The events necessary for cumulative selection are much too improbable to build a theory on. The events needed for the origin of life are even more improbable.

Dawkins did not make a convincing case for the spontaneous origin of life. He only described some unsuccessful attempts to solve the problem. But his failure to convince is not surprising. Even the army of research scientists working on the problem for the past generation have not succeeded.

———

Dawkins gave the results of some computer simulations he made of evolution. His simulations were featured in the Computer Recreations Department of the June 1988 issue of *Scientific American*. In his book Dawkins described two kinds of simulations. One of these was the evolution of the text of a line from Shakespeare. The other was the evolution of a set of line drawings. Let's examine his simulations and

---

[*] This procedure is approximately correct only if the result turns out to be much less than 1. Otherwise a somewhat more complicated calculation is required.

see if they can be taken as realistic. Let's see if we can conclude anything from them about the neo-Darwinian process.

In his first simulation, Dawkins started with a string of random letters and had it evolve toward a target string. The target string he chose was the line spoken by Hamlet, "Methinks it is like a weasel." This string has 28 characters, counting spaces but not counting the period. He chose the characters of his starting string at random, using only capital letters to keep things simple.

He copied his starting string a few times to get the strings of the second generation. Each time he copied the string, he introduced a fixed chance of a random error. Because of these errors, some of the strings of the second generation were not the same as the starting string. The reproductions were meant to simulate those of a living organism, and the errors were meant to simulate mutations.

From the offspring strings in the second generation the computer chose the one most nearly like the target sequence:

```
METHINKS IT IS LIKE A WEASEL
```

The computer's choice of the best match was meant to simulate natural selection. From the selected string, the computer bred the third generation just as it did the second. Errors were again introduced in the copying, and the string that best matched the target was chosen for breeding the fourth generation. He continued the process for as many generations as it took to achieve the target string.

He gave the following example of one of his simulations. The random starting string of 28 characters was

```
WDLTMNLT DTJBKWIRZREZLMQCO P
```

After 10 generations he got

```
MDLDMNLS ITJISWHRZREZ MECS P
```

after 20 generations

```
MELDINLS IT ISWPRKE Z WECSEL
```

after 30

```
METHINGS IT ISWLIKE B WECSEL
```
after 40
```
METHINKS IT IS LIKE I WEASEL
```
and at the 43-rd generation he got his target string
```
METHINKS IT IS LIKE A WEASEL
```

Dawkins ran the simulation several times and found that he needed from 40 to 65 generations to get to the target string. He also found he needed an average of only about 11 minutes of computer time per run.* He contrasted this performance with what he would have needed to get the same result without the selection.

Dawkins acknowledged that his model did not portray the exact way living things operate. He acknowledged that to some extent his model is "misleading in important ways." The one misleading feature he mentioned is that his criterion for selection is based on a long-range goal — namely to get the target sequence. The criterion of natural selection, on the other hand, must be short-range and immediate. The string selected at each stage must itself be adaptive.

We have already noted that one factor responsible for the slow pace of evolution is the low rate of mutation of a DNA symbol. I also noted that unless we want to rule out convergence, we must say that good mutations are rare. Moreover, I noted that when a good mutation does occur there is a large chance it will disappear because of random effects. These factors conspire to keep the neo-Darwinian theory from working.

Dawkins's simulation model is, of course, artificial and is not bound by the requirements of living organisms. The mutation rate in his simulation could be much higher than what is allowed in real life. In real life the genomes are much larger than what Dawkins used in his simulation. Therefore,

---

* With the faster computers of today, it would, of course, take much less time.

in real life, the mutation rate must be small so as not to lose what has already been built up.

The string of letters in Dawkins's simulation is much shorter than the genome of a living organism. His string can therefore tolerate a much larger mutation rate than can a living organism. Instead of a mutation rate of 1/10,000,000,000, as we find in nature, he could get away with a rate of 1/28. If all 28 characters of the sequence were already correct, this mutation rate would produce error-free sequences in about 36 per cent of the reproductions. If the sequence were only half correct, then this mutation rate would produce error-free sequences in the correct half about 60 per cent of the time. There is no guarantee in real life that one gets the best mutation rate for evolution. Dawkins's choice of a large mutation rate is one reason that his simulation works, but does not represent life. We, therefore, can't expect his simulation to give us any insight into how well evolution works.

There is another problem with Dawkins's simulation. We found in Chapter 4 that evolution had a chance of working only if Darwinian Assumptions 1 and 2 held. Because we found that those assumptions do not hold, we had to reject the neo-Darwinian theory.

In Dawkins's simulated world, he could do as he pleased. He made the Darwinian Assumptions hold:

1. The criterion Dawkins used for the selection in his simulation does not match that of the real world. As he noted himself, his criterion is based on a long-range goal. That's enough to ensure that the Darwinian Assumption 1 holds in his simulated world, as it does not in the real world. At any stage of the simulation, before the goal is reached, there are incorrect symbols in the string. A change of any of these symbols to its correct value would be a good mutation and would, therefore, be a candidate for selection. Therefore, at any stage before the goal, the Darwinian Assumption 1 is guaranteed to hold.

2. The mutation rate Dawkins used was high, and his selection was immediate and certain. He didn't need the Darwinian Assumption 2 to get evolution to work. In my example of horse evolution, on the other hand, the mutation rate was 10^{-10} per nucleotide per replication. Moreover, random effects would be likely to get rid of even a good mutation. Under those realistic conditions we found that there had to be a potential of a million adaptive mutations at each step to make the neo-Darwinian theory work. There had to be that much potential to ensure that at least one of them would occur and survive.

Dawkins used a high mutation rate, and a large fraction of his mutations were "good" ones. He was certain to have at least one good mutation at each step. Moreover, he selected in a way that ensured that a good mutation would survive. He didn't leave it to chance.

The same calculation I did on horse evolution can be used on Dawkins's simulation. Such a calculation shows that his simulation will work. He used a mutation rate of about 1/28 per character per replication. The average number of replications he needed to get to the target string was about 240. His results are consistent with this number.

If he had run a more realistic simulation he would have been at his computer for a long time. His simulation would have been more realistic had he used a genome of, say, 500 symbols instead of only 28, and a mutation rate of 10^{-10} instead of 0.04. Had he done that, he would have needed some seventy billion replications just to get the mutations he needed. He would need still more to make selection spread the 500 mutations through the population. Without simulating the selection, he ignores the likelihood that a good mutation will be wiped out before it takes over the population.

———

Dawkins's second simulation was the evolution of line drawings. He wrote a program to draw symmetric figures in a tree-like fashion, where new lines would branch off old lines recursively. The program for drawing the figures had a symmetry built into it. He used nine traits to define the drawings, and he gave the traits numerical values. They could take on the value of any integer from −9 to +9. The traits included the angles between the branches, the lengths of line segments, and the depth of the recursion.

He called his figures "biomorphs" after the biomorphs of Desmond Morris's surrealist paintings. This allowed him to write about them as if they were indeed a population of living organisms that were evolving.

There are nine traits and each can have any of 19 different values. Since each set of nine trait values defines a figure, there are 19^9 different possible figures in his repertoire. That's about 300 billion. He could start with any figure in his repertoire, and he could get to any other by his evolution program.

He likened the trait values to genes, since they are "heritable" and they determine the figure completely. Changes in a trait value play the role of mutations in a gene. The simulation copies a figure, adding the smallest possible mutation. That's a change of one unit in one of the trait values. There are two possible changes that can be made in each trait value at each step, namely, an increase or a decrease by one unit. There are 18 different smallest mutations, and there are therefore 18 different figures that could be made at any stage (except when the trait value is already at its limits of either −9 or +9).

He used the equivalent of a high mutation rate because, as he said, "humans haven't the patience to wait a million generations for a mutation!" At each step he generated all 18 nearest mutants, and he examined the looks of each. At each stage he selected one for further breeding. He repeated the process for as many steps as he had patience to play with it. The drawings were built as a succession of basic tree-like branches, but after the lines have crossed and recrossed several times the result often has little resemblance to a tree.

With the symmetry built into the figures, one can, as he looks at them, imagine they have some kind of form. One can see in some of them sketches that could look like insects and airplanes.

In the simulation, the operator (Dawkins himself) does the selecting. The simulation was of artificial, rather than natural, selection. By choosing figures that please his eye, he could make them develop in any one of many directions. Because the steps are small, figures tend to look like their parents. In this way the passage from one step to the next looks more or less continuous. Also, since the steps are small, the operator feels he has some control over the direction of the evolution. Dawkins offered this simulation as a second example to show that evolution can work.

But in this simulation, too, Dawkins did not capture the points that are important in deciding whether or not evolution can work. The simulation lacks two important features.

1. There are no lethal mutations for his figures. *Every* figure can survive and reproduce. Only the operator decides if a figure is to reproduce. There is then no limit to the mutation rate one can impose, allowing the simulation to be as fast as one wants.

2. There are very many adaptive mutations. Any figure can be adaptive. It depends only on the operator's choice.

In real life, too high a mutation rate is harmful to the organism. A high mutation rate could destroy the population.

One problem with evolution is that the chance of it happening in a reasonable time is too low. Or what amounts to the same thing, it takes too long. When Dawkins speeds up evolution enough, he is glossing over the very problem that makes it impossible. Cumulative selection doesn't work because the low mutation rate makes the chance too small of getting a good mutation. A low mutation rate means a long wait for a good mutation to occur. The simulation may cheat in the time only to save the human operator the inconvenience of waiting. But that throws out the very thing that keeps evolution from working in real life. For this reason alone, results from such a simulation can't be used to show that evolution works.

There is a second reason evolution works in the simulation even though it doesn't work in real life. At any stage in the simulation there are many mutations that can be selected. The operator picks the figures on the basis of how "interesting" they look. That means that almost any figure of more than just a few lines could be adaptive.

Let us see just what the operator does during the simulation. At each stage he sees the set of all figures that differ from the latest "breeding" by one unit in one of the trait values. These are "mutations", and there are 18 of them. The operator would usually want to go on with the simulation at any stage rather than end it. He may not want to end it even if he thinks at one point that he sees no really interesting figures. He will pick one of them anyway just so he can proceed.

But if we replaced the operator by an algorithm, the simulation might have a chance to fail. If the simulation would act like natural selection, it could fail. Dawkins's simulation has no chance to fail. If the operator were to follow fixed rules in his selection, he could reach a stage where none of the figures he saw would satisfy the criteria for selection. The evolutionary process would then stop. An operator will tend to avoid such blocks by making his rules flexible.

Beyond all the above, no one knows if organisms in real life can mutate through many steps, where in each the mutation is a change in a single base pair and is adaptive. No one knows if there can be chains of point mutations in living organisms where each can be adaptive in some environment. Moreover, there are *no examples* of mutations that add any information to the genome. Dawkins's simulations have nothing to do with organic evolution.

Thus, we see that the simulation of the line drawings cannot serve to demonstrate evolution. Because of the way it's built, the simulation sidesteps the reason evolution doesn't work in real life.

————

We've seen in Chapter 4 that cumulative selection doesn't work. Dawkins gave no analysis to support his belief that cumulative selection does work. Nor did he show that it can serve as an explanation of evolution.

Nor did his treatment of the origin of life make a convincing case for a natural origin. He had to deal with the origin of life if he wanted to make his case that life was not created by a Creator. Dawkins only presented the problem. He didn't solve it. He described one suggestion for the origin of life, but he gave no useful quantitative information for it.

Nor did Dawkins's simulations support his case for cumulative selection. In trying to make his simulations simple, he threw out the factors that keep cumulative selection from working in real life.

The neo-Darwinian theory is a poor basis for Dawkins's belief in the natural origin of life. That belief led him to believe that there is no Creator and no Ruler of the universe. That's a profound conclusion. It's a shame such a profound conclusion is based on such a poor theory and such flimsy evidence. But like many passionate believers, Dawkins did not examine his evidence critically. He let his heart lead his mind, even though he would like to think that he came to his conclusion in a rational and scientific way. The dust jacket of Dawkins's book says,

> There may be good reasons for belief in G-d, but the argument from design is not one of them.

I would put it differently:

> There may be good reasons for being an atheist, but the neo-Darwinian theory of evolution isn't one of them.

Chapter 7

THE DECK IS STACKED!
Nonrandom Variation

COWBOY Chuck had accused the dealer of stacking the deck because he suspected the deal wasn't random. Were you to say a deal isn't random, you would be accusing the dealer of stacking the deck. When evolutionists say genetic variation is random they mean to say the chance of a variation occurring has nothing to do with the way the variation helps the organism adapt to its environment. When I say a variation is not random, I mean that the chance of it occurring has something to do with how the organism adapts to its environment. I mean that the adaptation in some way has been influenced by the environment or the needs of the organism.

A half century ago, at about the time the neo-Darwinian theory was being formulated, the famous biologist D'Arcy Wentworth Thompson [1942] was criticizing the role of natural selection:

> We begin to see that it is in order to account not for the appearance but for the disappearance of such forms as these that natural selection must be invoked. And we then, I think, draw near to the conclusion ... that the great function of natural selection is not to originate but to remove. ... we ... see in natural selection an inexorable force whose function is not to create but to destroy — to weed, to prune, to cut down and to cast into the fire. [pp. 269-270]

On this he quoted Yves Delage [1903]:

> La sélection naturelle est un principe admirable et parfait-
> ment juste. Tout le monde est d'accord sur ce point. Mais
> où l'on n'est pas d'accord, c'est sur la limite de sa puis-
> sance et sur la question de savoir si elle peut engendrer des
> formes spécifique nouvelles. *Il semble bien démontré au-*
> *jourd'hui quelle ne le peut pas.** [p. 397]

The neo-Darwinian theory (NDT) argues back that natural
selection's power to create, lies in its very power to cull. Millions of new forms of all kinds are presented to it by the
powers of random variation. It selects from those that are
adaptive.

But we have concluded that the random variation described
by the NDT does not have the power to create enough nov-
elty to be selected. I am not the first to arrive at this conclusion. Several biologists and students of evolution have arrived at the same conclusion. Mae-Wan Ho of the Open University of Milton Keynes and Peter Saunders of the University of London have said [Ho and Saunders 1979]:

> ... the variations of the phenotype, on which natural selec-
> tion could act, do not arise at random ....

Random mutation has been described as "unprovable" and
"statistically unlikely" [Cook 1977], agreeing with our conclusion of Chapter 4.

The NDT, based on random variation, cannot account for
large evolutionary changes. It can account for only a limited
class of small changes. There is no evidence that random
mutation and natural selection played any role in the origin of
any of the major groups of organisms, including species [Ho
and Saunders 1979]. The NDT doesn't account for the origin
of the phyla, the classes, or the orders. It doesn't even account for the origin of species, except in special cases that we

---

* Natural selection is a laudable and entirely correct principle. Everyone is in agreement on this point. But where they are not in agreement
is on the limit of its power, and on whether or not it can produce specific new forms. *Today it seems well established that it cannot do so.*
[My translation, emphasis in original. *LMS*]

may well call trivial compared to the great sweep of life, and which are cases that cannot be extended to include macroevolution.

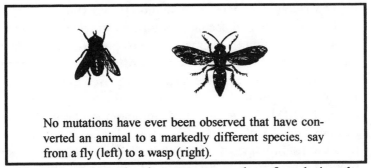

No mutations have ever been observed that have converted an animal to a markedly different species, say from a fly (left) to a wasp (right).

Neo-Darwinians do cite some examples of evolution they claim the NDT explains. Natural selection can indeed account for the replacement of light-colored moths with dark ones when the lichens get covered with soot [Kettlewell 1955, 1956, 1957, 1959, 1973]. But no one has shown that the difference between the two kinds of moths arose by a random mutation. The key point of randomness is not tested by those data.

But the NDT claims to account for much more than changes in moth-population frequencies. It claims to account for how the moths came into being in the first place. It claims to account for all evolutionary change and the origin of all living things. But this grand claim is not supported by evidence. Let's not confuse the little the theory may explain with all that it claims and with all the average layman has been led to think it explains. Natural selection, working on changes on a small scale, is only a faint shadow of what is claimed for the theory of evolution.

The dramatic claims made for the theory are what have attracted the world's attention to evolution. They are what have encouraged some prominent evolutionists to express opinions in the name of science that exceed the bounds of their expertise. They have prompted the late paleontologist and leading evolutionist George Gaylord Simpson to say that Darwin's theory has given man a new concept of himself [Simpson 1960]. They are what have prompted the prominent geneticist Bentley Glass to try to derive ethical values for mankind from the neo-Darwinian theory of evolution [Glass 1965].

The NDT claims to account for how all life developed from a simple beginning. It claims the development of life to be a natural phenomenon. Since we know of no physical laws that can make *organisms* evolve, Darwin proposed doing it statistically with *populations.**

But the NDT can account only for evolution brought about by some small isolated changes. The NDT does not explain how the information of life could have been built up. When proponents of the NDT try to use it to explain large changes, their argument becomes vague. At best, their argument becomes just a scenario of what *might* have happened.

When the NDT was established as the *modern synthetic theory of evolution* some 50 years ago, it described macroevolution as the result of a long series of small steps [Huxley 1943]. Their description followed Darwin's hypothesis of *accumulative selection.*

At first, a minority of evolutionists refused to accept a gradual process as the explanation of macroevolution. They saw no way that cumulative selection could achieve the host of auxiliary adaptations that have to accompany a large adaptive change. The paleontologist O. H. Schindewolf [1950] saw real gaps in the fossil record, and he could not accept a slow gradual evolution as its explanation. The geneticist Richard Goldschmidt [1940, 1952] proposed to account for the large evolutionary changes as the result of sudden large mutations, or *saltations.* The zoologist H. Graham Cannon [1958] suggested that a Lamarckian mechanism might be operating, and that adaptive variations may indeed

---

*Darwin didn't actually enter into the statistics of population genetics. This was done later by Sir Ronald Fisher and by Sewell Wright. They expressed Darwin's basic ideas in mathematical form.

have been induced by the environment. He did not, however, suggest a mechanism to account for how this might be done.

The neo-Darwinians were not persuaded by the arguments of the minority. They rejected large mutations as a way of getting large evolutionary steps. They invoked the small mutations, known to produce microevolution, to account for macroevolution as well [Rensch 1954, 1960; Simpson 1953]. They disdained the large mutations because they are almost always harmful. They are almost never of any good. Large adaptive mutations are, in principle, too rare to lead to evolutionary change even with large populations and over long time periods.

Most evolutionists today hold the variations on which evolution is based to be random, but they are divided on how to explain macroevolution. The majority are the ones I have called the *bradys*. They hold with the founders of neo-Darwinism and say that evolution proceeds slowly. They hold that large-scale changes, which are what Darwin set out to explain, are made up of long strings of many small ones. As previously noted, some of the punctuationists also belong in this group. But no one has shown that small random changes can build up to make the large ones. Nor are they able to account for the information that, according to them, must have been built up in living things.

The bradys must hold that, on the average, cumulative selection has to add a little information to the genome at each step. But of all the mutations studied since genetics became a science, *not a single one* has been found that adds a little information. It is not impossible, in principle, for a mutation to add a little information, but it is improbable.

The NDT was an attractive theory. Unfortunately, it is based on the false speculation that many small random mutations could build up to large evolutionary changes. In Darwin's day there was no scientific evidence for or against such a speculation. Nor did any evidence for it arise during the eighty years that elapsed from the publication of Darwin's book until the NDT was set up. The speculation was never-theless accepted as possible, even as fact. But during the half century of the NDT, we have probed the molecular level of cellular functions. Now, as we come to the close of the twen-tieth century, we have a lot of evidence of the nature of the mutations to which the neo-Darwinians assigned the role of the small variations. None of this evidence vindicates the Darwinian speculation that large-scale evolution has its source in random variation. All evidence is against it.

Those I have called the *tachys* oppose the bradys' view. The tachys are the successors of those who objected to the gradualness of the NDT forty and fifty years ago. They fol-low those earlier objectors in rejecting the long slow process of cumulative selection and follow Goldschmidt in proposing that large changes can come about suddenly [Stanley 1979]. But they say that we now know of molecular and genetic mechanisms that might account for sudden large changes [McDonald 1990]. Large evolutionary changes, they say, may have come about through genetic rearrangements in regula-tory genes. A change in only the timing of a molecular signal during embryonic development, for example, could result in a large change in the phenotype [Gould 1977].

Yet, most tachys endorse randomness. But if their varia-tions are random, they too cannot account for a build up of genetic information. The chances are almost nil that a large chunk of information can get into the genome in this way.

That's why the neo-Darwinians chose to base their theory on a long succession of small changes. Since it has lately be-come clear that such a theory can't work, some of the succes-sors of the neo-Darwinians are trying to save the theory by interpreting the variation as large changes in regulatory genes. They are, though, maintaining the randomness. Yet the neo-Darwinians were right in rejecting large changes. You cannot expect to get a large adaptive genetic rearrangement by chance. Genetic rearrangements seem to be a normal ac-tivity of the cell. Genetic rearrangements seem to be as normal an activity of the cell as cell division, even though they don't happen so often. They are mediated by specific enzymes which, as far as we know, the cell synthesizes just for that purpose. Some biologist may want to call them random only because their timing is not yet understood.

Some want to say an organism can gain information if it assimilates a piece of DNA from another organism. It can. But if we want to know how the information has been built up in living things from a simple beginning, we must address the question of how that other organism got the information. Trying to build information by transferring segments of DNA from species to species is like trying to build an economy by having everyone take in each other's laundry. I'm reminded here of a story I heard from Herman Kahn.

> The boy's father smiled when his son told him his dog was worth $100,000. He laughed outright when the son said he had a customer for it. The next evening the father asked his son if he had made the sale. "Yep," said the son. "Did you get your price?" the father asked condescendingly. "Yep." "Really?" asked the father incredulously. "You got $100,000 for your dog?" "Yep. I traded him for two $50,000 cats!"

For an economy to grow, wealth must be created. For information to build up in living organisms, it must be created somewhere.

If I don't believe the poker hands dealt out are random, I have to look for the cause. Did the dealer fiddle with the deck? If the variation leading to large-scale evolution cannot be random, we must look to nonrandom sources for the variation. There is a large body of observations showing that the deck of life is indeed stacked.

Observed adaptations induced by the environment have been reported for more than a century, but they have been largely ignored by mainstream biology due to the dominance of Darwinian theory and the NDT. They have been swept under the rug because they couldn't be understood within the framework of the theory. Some geneticists have felt compelled to apologize for observations of this kind. They discounted them as "unfortunate defects in the delicate genetic apparatus".[*] Sir Ronald Fisher has been reported to have said that it is not surprising that such elaborate machinery should sometimes go wrong [West-Eberhard 1989].

In the past two decades some biologists have begun to suggest that the variation driving evolution may indeed not be random [Ho and Saunders 1979; Cook 1977; Rosen and Buth 1980]. The variation is triggered by the environment to which the organism will then adapt. Cook [1977] has noted that the NDT is on weak ground both mathematically and biochemically, and he expressed serious doubt about the usefulness of random variation in evolution.

Only a minority of biologists today hold that the variation of evolution cannot be random. In this point the minority are on more solid ground than the majority. If we want to gain a true insight into evolution, we must look to nonrandom variation.

———

Living organisms respond to their environment on several levels. To survive and live well, an organism senses its environment and adjusts itself accordingly. We have seen one level of this adjustment in Chapter 2, in the genetic control of enzyme synthesis in a cell. This genetic control system senses, for example, the presence of the enzyme's substrate and turns ON the gene that encodes the enzyme. The system turns genes ON or OFF as they are needed, but makes no permanent change in the genome. This kind of control permits the organism to operate efficiently under a variety of conditions, enabling the organism to live efficiently through *short-term* changes in the environment.

---

[*] See West-Eberhard [1989] and her comments on this.

An extension of such controls can lead to changes on the long-term, or evolutionary, level. Moreover, if the control is in the development process, even a small change of the "right" kind can lead to a large adaptive change in the phenotype. But the "right" kind of change would be unlikely to occur by chance. If the changes are random, the chance of a "right" change occurring is proportional to the fraction of "right" changes among all possible ones. The number of "wrong" changes is vastly greater than the number of "right" ones. But — and here is the important point — if the genome were *set up* for an adaptive change to be triggered by a cue from the environment, then chance wouldn't be involved. The right adaptive change would be sure to occur when it was needed.

———

There are several different kinds of variations of the phenotype that can be induced by the environment, and many of them can lead to long-term changes in a population. These variations can be divided into two broad classes. In the first class are variations in the phenotype that result from changes in the DNA sequence. In the second class are variations in the phenotype without a change in the DNA sequence.

Nonrandom mutations fall into the first class. Mutations, as we have seen in Chapter 2, are changes in the DNA sequence, and can be divided into several types. Some are changes in a single nucleotide, and some are more complex. Many mutations are known to be spontaneous and their effects are independent of any environmental influence. These are not the mutations of macroevolution. Evidence indicates that these mutations are the results of errors in the proper working of the genetic mechanism. Most of them result from errors in the replication of the DNA.

The mutations I am calling for are those that show evidence of being nonrandom in that they are triggered by the environment. Some of them have been seen to be adaptive. These mutations form the first class of nonrandom variation that could lead to observed evolution. These mutations may act as switches triggered by the environment that switch the genome to one of a preexisting set of potential states to produce an adaptive phenotype.

We have seen that random mutations do not put information into the genome. The mutations that contribute to macro-evolution are nonrandom — they are triggered by the environment and lead to adaptive phenotypes.[*] The potential for adaptivity to the environment already exists in the genome. The environment just triggers it.

Not all mutations are random. Some are known to be non-random in time, some are known to be nonrandom in place. Some occur just when they are needed, and some occur in just the right place on the genome where they can be effective. Many mutations, such as those described below, are executed with precision and are under elaborate cellular control. To dismiss them as random would be to turn a deaf ear to what nature is telling us.

My suggestion here is speculative. We do not yet have an example of a mutation that is sufficiently well understood on the molecular level that we can see (1) just how the environment triggers it and (2) how it produces a phenotype adaptive to that same environment. But we have recently discovered the molecular nature of some adaptive mutations that are triggered by the environment. Although at this writing we do not yet know the molecular mechanics of just how the environment does the triggering, we do know that the triggering occurs.

One well-known type of mutation that does not seem to be entirely random is an *inversion* of a section of the DNA (see Chapter 2). An inversion within a gene can garble it and make it inactive. If the inversion is later precisely reversed, the gene will recover its activity. Alternatively, an inversion

---

[*] Biological mechanisms, just like computer programs, sometimes perform poorly in the event of unusual inputs. In such cases, a change in the phenotype induced by the environment will not be adaptive.

can remove a promoter from an operon and shut down an entire battery of related genes. Precise inversions can thus serve either to stop specific genetic activity or to restore it. An inversion can in this way produce a sudden large heritable modification in the phenotype [Darnell et al. 1986]. Inversion patterns in chromosomes have been used to characterize some species of *Drosophila* [Dobzhansky 1951]. It is likely that the inversions themselves were the genetic change that led to the formation of the species.

An interesting example of an inversion is one that's known to occur regularly in the bacterium *Salmonella*. This mutation results in replacing one type of protein in the flagella with another [Silverman and Simon 1983, Darnell 1986 pp. 441-42, Griffiths et al. 1993 pp. 572-73]. These two proteins are, as far as we know, identical in function. They differ only in their antigenic properties.

How this inversion occurs is well understood, as is its effect on the making of the flagellin proteins. The genes that encode these proteins can be in either of two states. In one state, one of the proteins is made; in the other state, the other is made. A specific and precise inversion makes the genome pass from one state to the other. A segment of the genome of about 1000 nucleotides seems to be assigned specifically to the inversion. The segment has special short DNA sequences of 14 nucleotides at each end that permit the inversion to occur through a recombination. A protein necessary for the recombination is encoded within the 1000-nucleotide section. The special markings at the ends of the section ensure that the segment will be inverted precisely, and that it will be precisely restored by another inversion. The inversion acts as a switch to change the flagellin protein.

So far as we know, these inversions occur more or less regularly. They occur about once every 10 generations. Periodic switching from one antigen to the other helps the bacterium evade its host's immune system. The state of the genome is heritable. The antigen not being attacked by the immune system will tend to increase its numbers.

Recombination in the genome can also produce other kinds of mutations. It can produce deletions, duplications, and translocations. Recombination is not a random process. It is under strict genetic control, and requires several special enzymes for its operation. There are even special genes that affect the efficiency of the recombination [Griffiths et al. 1993, pp. 571 ff.].

———

Genetic elements have been discovered that move from place to place on a chromosome. They even move from one chromosome to another. These are short pieces of DNA found in bacteria, plants, and insects, and are called *transposable genetic elements*. Geneticists believe that these elements exist in all organisms. Two kinds of such elements are recognized. The smaller ones, having up to about 1,500 nucleotides, are called *insertion sequences* and are denoted by the initials IS. The larger ones, *transposons*, have 2,500 nucleotides or more. Some of them have as many as 20,000.

Insertion sequences can, as far as we know, move into any part of the genome. If they fall within a gene, they generally garble that gene and make it ineffective. An IS can contain transcription-termination signals, in which case it will turn OFF not only the gene it lands on, but it will also turn OFF any downstream genes in the same operon. The same IS's appear in different organisms, and they have been denoted by *IS* followed by a number, such as IS1 and IS5.

Since transposons are larger than IS's, they carry more genetic information. They are denoted by Tn followed by a number, such as Tn1 and Tn903. The more notorious ones in bacteria carry genes for resistance to several kinds of antibiotics. A transposon carrying resistance to ampicillin (Tn1) has about 5,000 nucleotides. That carrying resistance to tetracycline (Tn10) is about 10,000 nucleotides long. The one carrying resistance to ampicillin, streptomycin, and sulfanilamide (Tn4) is about 20,000 long. They can move from a plasmid of the bacterium into its DNA. They can also move from one bacterium to another during conjugation.

Barbara McClintock, who received the Nobel Prize in 1983 for her work on genetic rearrangements, noted that there are indications that these genetic modifications occur in response to stress [McClintock 1978]. Exactly what their normal function is, aside from spreading antibiotic resistance, we do not yet know.

Barry Wanner of Emory University has suggested that genomic rearrangements could be part of a control system in bacteria that would produce heritable changes in response to environmental cues [Wanner 1985]. Virginia Walbot of Stanford University and Christopher Cullis, later at Case Western Reserve University, have noted that genetic rearrangement can be responsible for rapid genetic changes, especially in plants [Walbot and Cullis 1985].

Over a decade ago Barry Hall, then at the University of Connecticut, prepared a strain of *E. coli* bacteria that could not break down the milk sugar, lactose [Hall 1982]. Normal *E. coli* can live on lactose because they have an array of enzymes that can metabolize it. For this set of experiments Hall prepared a strain that lacked the gene encoding the first enzyme of the array. Because of this lack, his strain of bacteria could not live on lactose. When these bacteria grew and multiplied on another nutrient, but in the presence of lactose, *two mutations* were found to appear in the same bacterium. One of these mutations was in a hitherto unknown structural gene and the other was in its control gene. The mutated structural gene encodes an enzyme that can perform the missing first step in lactose metabolism. Because the mutant bacteria activated a substitute gene, it was able to live on lactose. The gene that mutated had been present all along, but it was dormant. Its normal function is unknown. Hall called it a "cryptic" gene.

Neither of the above two mutations is of any use by itself to the bacterium. For the bacterium to metabolize lactose, both mutations have to occur. In the absence of lactose, these two mutations are independent. They will occur together only by chance, and will do so only with the small probability of only 10^{-18}. Hall calculated for his population the expected waiting time for both these mutations to occur by chance. He found that if they really did occur at random, he would have to wait about *a hundred thousand years* before he could expect to see one of these double mutations. But in the presence of lactose he found about 40 of them in just a few days! These results suggest that the lactose in the environment induced these mutations.

Hall [1988] did another experiment with the same kind of result. He grew bacteria in the presence of salicin, which is a nutrient that the bacteria cannot normally break down and use. But the surprise was that the bacteria have another "cryptic" gene, normally repressed by a regulatory gene, encoding an enzyme that can break down salicin. The cryptic gene will become active if one of a few mutations would occur in its regulatory gene. One is a specific nucleotide substitution. Another is the insertion of either one of two transposable elements.

He used a strain of bacteria whose cryptic gene itself did not work. His strain had in its genome an extra piece of DNA, an insertion sequence known as IS103, having 1,400 nucleotides. It sits upstream of the cryptic gene, and keeps it OFF because it shifts the coding frame and garbles the transcription from DNA to mRNA.

For Hall's strain to metabolize salicin, two genetic changes had to occur. The sequence IS103 had to be precisely deleted. Then the right nucleotide had to be changed, or else a sequence called IS1, or another one called IS5, had to be inserted into the cryptic regulatory gene. Hall tried to measure the spontaneous rate of the precise deletion of IS103. He found it to be too low to measure. He could say only that its probability is less than 2×10^{-12}, or about 2 chances in a trillion. In the absence of salicin these two mutations would occur in the same cell with a chance of less than 10^{-19}.

The NDT does not admit that the environment can cause an adaptive mutation to occur The mutations important for the NDT are said to be *random*, and their occurrence is supposed to be independent of each other and of the environment. If the two are indeed independent, then the chance of the right double mutation in at least one cell of the population in two weeks is about one in thirty million. Hall would have had to wait about a million years to see one of them in his culture. Yet he found that within two weeks about *60 per cent* of his colonies underwent both mutations and could metabolize salicin! He investigated in particular the occurrence of the precise deletion of IS103 in the presence of salicin. (Remember that in the absence of salicin the rate of this deletion is too low to be measured.) In the presence of salicin, he found that within 8 to 12 days as many as 89 per cent of the cells in a colony underwent the deletion!

John Cairns and his team at the School of Public Health at Harvard University described another experiment with bacteria [Cairns et al. 1988]. They used a strain of bacteria with a defect in the gene encoding the enzyme for breaking down lactose. Their bacteria could therefore not live on lactose. Cairns's team fed their bacteria lactose and looked for cells to appear that could live on it. They found such cells, and they interpreted their appearance to be the result of mutations. Using the statistical approach established by Luria and Delbrück [1943], they concluded that, in the presence of lactose, mutations were occurring that convert a gene into one that makes an enzyme that breaks down lactose. They concluded that:

> The cells may have mechanisms for choosing which mutations will occur. ... Bacteria apparently have an extensive armory of such 'cryptic' genes that can be called upon for the metabolism of unusual substrates. The mechanism of

activation varies. ... *E. coli* turns out to have a cryptic gene
that it can call upon to hydrolyse lactose if the usual gene
for this purpose has been deleted. The activation of [this
cryptic gene] requires at least two mutations. ... That such
events ever occur seems almost unbelievable... [Cairns et
al. 1988]

These, and other similar, reports have stirred up much interest as well as controversy among scientists engaged in research in molecular genetics and evolution. When these experiments were first reported, they were met with skepticism.
Their results bring into question the status of the principle of
the independence of mutations from the environment [Stahl
1988, Benson 1988].

If the results of these experiments indicate that adaptive
mutations are stimulated by the environment, they contradict
the basic dogma of neo-Darwinism. According to that
dogma, mutations are random, and the kind of mutations that
occur are independent of the environment. If mutations are
really nonrandom in the sense that the environment can
stimulate adaptive mutations, then the paradigm of Darwinian evolution, which has dominated the biological sciences
for close to 150 years, must be replaced. In science, a paradigm of such stature cannot be allowed to fall easily. Experimental results that call for its fall must be examined and reexamined for all possible artifacts and defects. If there is any
possible alternative explanation of the experimental results
that can be consistent with it, no matter how far-fetched, the
paradigm will be retained. This conservatism is necessary to
prevent science from fitfully chasing after every harebrained
experiment.

Several molecular geneticists have suggested how the
above phenomena could be explained as the result of the
same kind of random mutations called for by the NDT. Some
scientists have suggested Cairns's results may indeed indicate
the failure of the principle of the independence of mutations
from the environment, while others have suggested that one

might find an explanation of these phenomena that leave the principle intact.

Molecular geneticists went to their laboratories and tried to test these suggestions by experiment. The experiments have not resolved the controversy, but rather have deepened it. Richard Lenski of Michigan State University, citing his own and other experiments, contends that artifacts could account for results that look like directed mutations [Sniegowski & Lenski 1995]. On the other hand, Patricia Foster of Boston University citing her own experiments and those of others, contends equally strongly that the effects are real [Foster 1992]. In the opinion of Sniegowski & Lenski [1995] the interpretation of Cairns's results is not yet resolved.

More recent experiments have revealed another kind of mutation that appears to be adaptive and brought on by the environment. These experiments show that, in the presence of lactose, adaptive mutations activating a dormant gene encoding an enzyme that will hydrolyze lactose in *E. coli* are of a kind different from mutations that occur in the absence of lactose [Foster and Trimarchi 1994; Rosenberg et al. 1994; Galitski and Roth 1995; Radicella et al. 1995]. See also Culotta [1994] and Shapiro [1995].

Darwinian evolutionists see the nonrandom interpretation of these experimental results as obviously incorrect because they contradict the neo-Darwinian dogma. I, on the other hand, see this interpretation as confirming, on the bacterial level, the nonrandom variation indicated by many examples in plants and animals — examples that Darwinian evolutionists have largely ignored because they do not fit in. Resistance to the nonrandom-variation interpretation stems from a refusal to abandon the Darwinian agenda that evolution must confirm that life arose and developed spontaneously. With that agenda, nonrandom adaptive variation, arising from an environmental signal turning ON an already present set of genes, is hard to account for. One's tendency to accept the nonrandom interpretation of the experiments on bacterial evolution depends on how strongly one insists on the necessity of the Darwinian agenda. Which interpretation is correct will ultimately be decided in the laboratory.

The several examples cited above indicate that the phenomenon, if it is indeed vindicated, may be widespread in bacteria. Just as these bacteria contain "cryptic" genes which encode for enzymes that are needed in some environments, so I suggest that other organisms also may have latent parts of their genome dedicated to be adaptive to a certain set of environmental conditions that may arise. The environment can then supply a cue that will turn ON the latent section that will make the organism adaptive.

Mechanisms are indeed known whereby sophisticated and elaborate environmental cueing can be achieved in animals. First of all, we know that the environment can act on the genes of the germ cells through messenger molecules known as hormones. We know the environment can stimulate the production of hormones, and we know that hormones can play a role in regulating genes in targeted cells. These are all the elements needed for my suggestion that genetic changes can be triggered by the environment.

The second class of nonrandom variation that could lead to evolution are those that lead to a change in phenotype, but do not change the DNA sequence. What kind of variations can these be? How can they be controlled by the environment? Moreover, how can they be heritable?

Genetic states that dictate the current metabolic activity of the cell through enzyme synthesis are usually not heritable. A gene that is normally OFF will be ON only so long as the signal that turns it ON is still present. Take that signal away and the gene goes OFF. The ON/OFF state of such a gene is generally not heritable. The state depends on the presence of its inducing or repressing signals.

But there are some genetic states that *are* heritable. The most outstanding examples of heritable genetic states are the changes in the genetic program that occur during the development of an embryo. During development, genes get turned ON and OFF as the cells divide. The ON/OFF state is passed from mother to daughter cell as the cells differentiate. Not any method of turning genes ON and OFF lends itself to being passed on through cell division to later cell generations. How cells during development pass on their genetic state to daughter cells is not yet well understood.

One known way of setting the genetic state and making it heritable is by attaching a methyl group (CH_3) to one of the carbon atoms of the cytosine bases in the DNA. Methylation serves to keep the gene OFF by not letting the regulatory proteins get to it. When the cell wants to turn ON the genes in a segment of DNA, it removes the methyl groups. Some geneticists hold that methylation may be one of the ways the organism might control gene activity during development [Holliday 1989]. The pattern of methylation is made heritable through an enzyme that acts during DNA replication. The enzyme copies the methylation pattern from the template strand of DNA onto the daughter strand as it is being constructed [Darnell et al. 1986].

Another way of making a genetic state heritable is to have the gene turn ON and OFF with a locking trigger. A trigger that turns a gene ON can lock it in that state if the gene itself causes the synthesis of the control protein that keeps it ON. Once the gene is turned ON it will remain ON even after the trigger is removed. Such a state can be heritable.

A locking trigger operates with *positive feedback*. The kind of genetic control we have seen so far is built on *negative feedback*. Negative-feedback control is ideal for holding a gene's activity at an arbitrary level set by the need for its product. The state of the gene under this kind of control is not heritable. It does not get passed on to the next generation.

Positive feedback, however, *can* lead to a heritable state. Such a locking trigger is a well-known phenomenon discovered in the infection of bacteria by lambda phage [Herskowitz and Hagen 1980]. Electronic devices often use positive feedback to drive themselves to either of two extremes: to turn OFF, or to saturate. It's the kind of control often used to make an electronic switch with two stable states, sometimes called a *flip flop*. A trigger will change the state of such a switch. The new state will lock itself in place and endure even after the trigger is removed. Only when the switch gets the opposite trigger will it flip back to the other state. The active memory of every computer uses positive feedback. An electronic switch of this sort will produce at its output either a 1 or a 0, and it will hold that value indefinitely until it is switched.

Appendix J shows how a genetic switch with a locking trigger is heritable. A cue from the environment can trigger the switch, turning a gene ON or OFF. The ON/OFF state of the gene will maintain itself even through cell division. The genes of the daughter cells will have the same state as the mother cell. The heritability will maintain itself indefinitely through any number of generations. A new trigger from another environmental cue can reset the gene.

———

There is another kind of variation that does not involve the genome at all, and is therefore not heritable. Yet it can produce what looks like evolution. Indeed, the results look so much like evolution that for all we know some of the best examples of evolution may be due to this nonheritable kind of variation.

How can a variation that is not heritable produce anything that looks like evolution? Timothy Johnston and Gilbert Gottlieb, of the University of North Carolina, have described how this can be. Moreover, they have noted that it is happening all the time, has been observed, is well known, and is well documented [Johnston and Gottlieb 1990].

Everyone agrees that an animal embryo develops under the joint influence of its genetic program and the environment. The same genome under different environmental conditions will often produce a different phenotype. Signals originating in the environment are no less influential in the developing embryo than are the signals originating within the organism itself. Indeed, one cannot entirely separate the two effects because they act only together [Johnston and Gottlieb 1990]. Signals from both sources act together in the development process. Different cells in an embryo, even though they have identical DNA, take different development pathways because their signal inputs are different. Indeed, that's what *differentiation* in the embryo is all about. The developmental pathway of the entire embryo would be different if some of the environmental signals were different.

Johnston and Gottlieb have gone so far as to claim that the distinction between *inherited* and *acquired* characteristics is not an operationally valid distinction. Genetic and environmental causes act together in development. For example, the adult forms of animal bone-and-muscle systems are to a large extent influenced by the forces exerted on them while they are growing. The bone-and-muscle system of the jaw, as well as the teeth, have to support the forces acting when the animal chews its food. The strength and direction of these forces depend on the kind of food the animal eats. The food may be hard or soft, chewy or crunchy, and it may come in large or small pieces. It is well known that the forces exerted on growing teeth, bones, and muscles will influence the forms these structures take on in the adult.

Rodents will readily sample new foods, and when they find something they like they make it part of their diet [Kalat 1985; Richter 1947; Johnston and Gottlieb 1990]. Young rodents get their taste for food from their parents [Galef 1985; Johnston and Gottlieb 1990]. A new seed, if it's tasty and nutritious, will spread through the population over cultural channels. If the seed remains available for a long enough time, it can become a standard food for future generations. The heritability of rodents' choice in food is perhaps as much cultural as it is genetic. Indeed, it may be more cultural than genetic.

The type of food a young rodent eats influences the adult phenotype. A new seed that's large and hard, for example, affects the development of the bone-and-muscle system of the rodent jaw [Frost 1973; Herring and Lakars 1981; Lanyon 1980; Johnston and Gottlieb 1990]. If a population of rodents were to shift their diet abruptly to a large hard seed, the phenotype of the next generation would change abruptly. The phenotype would change, but the genotype would not. The fossil record of these rodents would show an abrupt change in the jaw and tooth structure. Yet there would have been *no mutation and no selection.* The entire population would have changed together with the environment.

A change in phenotype in the fossil record is recognized as evolution. There is no way to tell from the fossils whether the observed changes in continuous records were caused by variation appearing in the genotype or only in the phenotype.

The influence of the environment on the form of fossils is a phenomenon well known to paleontologists. G. Fryer, of the Windemere Laboratory in Wales, and P. H. Greenwood and J. F. Peake of the British Museum [Fryer, Greenwood and Peake 1983, 1985] criticized the conclusions of the Harvard paleontologist Peter Williamson [1981a, 1981b, 1982, 1983, 1985a, 1985b] on just these grounds. They suggested that the changes in the shell structure of the snails he reported in the 400 meters of fossil record he studied in the Turkana Basin in Northern Kenya might not be the results of genetic mutations as Williamson proposed, but may have been only changes in the phenotype brought about solely by changes in the environment.

One can't help wondering how much of the fossil record might be the result of the direct influence of environment on the phenotype without any change in the genotype. We know the form and shape of teeth is strongly influenced by diet. Similarly, the form of bones is strongly influenced by the forces to which they are subjected during growth. Many of the fossils that have made the news as "missing links" consist mostly of teeth. Most of the rest are bones. What kind of support can fossils of bones and teeth, then, give to the randomness postulate of the neo-Darwinian theory?

Simpson [1961], in his story of horse evolution, has told us that when the great forests gave way to the great grassy plains, *Mesohippus* evolved into *Merychippus*. Through random variation and natural selection, he said, the horse developed harder, high-crown, teeth and evolved from a browser to a grazer to exploit the abrasive grass as a new source of food. He said:

> These horses were learning to eat grass and acquiring the sort of teeth that enabled them to do so. They were changing over from browsing animals to grazing animals ... It is not likely to be a coincidence that at the same time grass became common, as judged by fossil grass seeds in the rocks. [p. 173]

It may not have been a coincidence. But the change in the horses' teeth may not have happened the way Simpson described. The hard surface and the high crown of the *Merychippus* tooth may have been just the direct result of the horses eating the abrasive grass. Random mutation and natural selection may have played no role in the change from *Mesohippus* to *Merychippus*. We don't know how it really happened because we can't examine the genomes of these extinct animals. For many of them, all we have is a few fossil teeth. In any case, we see that what has been recognized as clear examples of evolution could have been the result of the environment acting on the developing embryo. It may have nothing at all to do with genetic variation or natural selection.

According to the picture I have given of macroevolution, we would expect that even a large evolutionary change would carry with it only a small genetic change. Even if the variation causing the change were genetic, we would not expect a

large change in the DNA sequence. If the second type of genetic variation I described were responsible for the change, there would be no change at all in the DNA sequence. If the first type were responsible, there would be some change in the sequence, but it would be small. In either case, the change in the DNA that caused the change in phenotype would not be large. On the other hand, if macroevolution went according to the NDT, then large evolutionary changes should come with large DNA changes.

John McDonald of the University of Georgia has pointed out the lack of correlation between the sizes of phenotypic change and DNA change. Differences in DNA between species seem to be unrelated to their supposed evolutionary divergence [McDonald 1990]. Citing the work of Allan Wilson and his co-workers [Wilson et al. 1974], McDonald noted the differences and similarities between frogs and mammals. There are two frog species, which are very much alike, but differ greatly in their genomes. The mammals, however, which have great differences in phenotype, differ little in genotype. These data indicate that the size of genetic changes may be unrelated to the size of phenotypic changes. Indeed, much genetic change may be irrelevant to evolutionary change.

As seen in Chapter 5, there are no known examples of random mutations that add information to the genome, as the NDT must require. Although in some special cases a loss of information can lead to an advantage for the organism, the large-scale evolution for which the NDT is supposed to account cannot be based on such mutations.

On the other hand, many examples have been reported of adaptive phenotypic changes triggered by environmental cues. (For starters see West-Eberhard [1986, 1989], Bradshaw [1965], Harrison [1980], Schlichting [1986], Stearns [1989], and the hundreds of references in these papers.) These examples fall into all the above categories. Some of them entail changes in the DNA sequence; some do not. But for most of them we don't know because they haven't been studied at the molecular level.

These examples of changes are nonrandom — the same environmental cue will elicit the same phenotypic change time after time.* The changes in the phenotype are more like the purposeful throwing of a switch than a random mutation. Living organisms have a wide complement of latent capabilities that can be turned ON by such cues.

Among the adaptations observed to be caused by a cue from the environment are those known as a change of *phase*. These adaptations can be changes in the structure of an organ, in its function, or in an animal's behavior [West-Eberhard 1989]. An organism's ability to change as the environment changes is known as *phenotypic plasticity*. It has been widely observed in both plants and animals for more than a century.

Plants and animals hardly ever increase their numbers so much that they exhaust their resources. They stop increasing well before they deplete their habitat. Animals are mobile and can disperse when they are overcrowded, but plants require other strategies. Some plants adjust their seed production to their density. It has been discovered that if plants are set close together, they will produce less seeds than if they are set further apart. Linseed plants, for example, have been reported to produce almost sixteen times as many seeds per plant when they are set far apart as when they are set close together. Other plants react to variations of density by varying the

---

* An environmental cue may be capable of eliciting any one of a set of variations, each leading to a phenotype adapted to a slightly different environment. Natural selection would then "fine tune" the population to the specific environment, whereas the cue would provide the "coarse tuning." In this case, the *set* of possible variations would be nonrandom, whereas the specific variation within the set would occur by chance. But since all variations within the set would appear in the population, the final result would be nonrandom.

numbers of their leaves or the lengths of their stems [Bradshaw 1965].

A Mediterranean grass has been reported to increase its flowering by a factor of a hundred when it was moved from less fertile to more fertile ground [Bradshaw 1965]. Several species of plants vary their stem height, stem number, and flowering time as conditions vary from sunshine to shade and from wet to dry [Bradshaw 1965].

Crabs prey on snails with thin shells, but they cannot eat snails that have thick shells. Snails can somehow tell if crabs are around. In the presence of crabs they grow a thick shell [Stearns 1989]. This adaptation clearly helps protect the snails from the crabs.

Snails are themselves predators. They prey on barnacles. When the barnacle senses snails, it protects itself by growing into a bent-over shape that keeps the snails from eating it. When there are no snails around, the barnacle develops into its normal straight form [Stearns 1989, Lively 1986].

I am suggesting here that organisms have a built-in capability of adapting to their environment. I am suggesting that to the extent that evolution occurs, it occurs at the level of the *organism*. This suggestion differs sharply from the thesis of the NDT, which holds that evolution occurs only at the level of the *population*. Organisms contain within themselves the information that enables them to develop a phenotype adaptive to a variety of environments. The adaptation can occur by a change in the genome through a genetic change triggered by the environment, or it can occur without any genetic change.

This capability is an advantage to the population. It helps it survive by adapting to long-term changes in the environment. Suppose a "biological engineer" is designing living organisms. If he has to design a species to live in a range of environments, he is faced with a design dilemma. On one horn of the dilemma, he could make the species well adapted and highly specific to one of the environments. But then it would be less well adapted to other environments. On the other horn, he could make the species less specific and broadly adapted to the whole range of expected environments. But then it would not be so well adapted to any of them as it could have been to one of them. A good way for him to resolve the dilemma would be to build into the species the ability to switch among several forms, each highly adapted to one of the environments. He would design the switch to be triggered by a cue from the environment. I am suggesting here that living organisms have the capability of switching from one phenotype to another when cued by the environment.

For the organism to have this capability, it has to have in it the necessary information. No one yet knows how much capability of this sort is built into free-living cells, plants, and animals. The more built-in potential there is, the more the information the organism must carry. One would expect the seat of this information to lie in the genome. Can the genome carry all this information? There is a vast amount of DNA in plants and animals whose function is as yet unknown.* Could the role of some of this DNA be to encode the potential phe-

---

* Some part of this is surely the encoding of the "cryptic genes" discovered in bacteria.

notypic diversity needed to adapt to a variety of environments? Mitochondrial DNA[*] and plastid DNA[†] also may play a role in coding this information. Indeed, both mitochondrial DNA and plastid DNA have been reported to have some effect on the phenotype in plants [Walbot and Cullis 1985].

———

Finches are the family of birds we see most. They comprise the largest family (Fringillidae) of the largest order (Passeriformes) of the class of birds. In many parts of the world there are more finches than there are other birds, both in the number of individuals and in the number of their species. Also known as sparrows and warblers, the finches are the most diversified of all bird families. Their evolution has been studied more than any other bird family.

The Galápagos Islands lie isolated in the Pacific Ocean. They straddle the equator, and are about 650 miles west of mainland South America. Darwin visited these Islands on his famous voyage on the H.M.S. *Beagle*. Among the birds on the islands, he found finches that looked like finches elsewhere, but they were not the same as any he had seen before. He collected samples of nine species of them and brought them back with him to England. There John Gould, an experienced systematist, examined the specimens and declared the birds to be different from all known species. Darwin theorized that some time in the past, a few finches found their way to the Galápagos Islands from the mainland. He suggested that since then, variations have appeared in the birds, and these changes were subject to natural selection. As a re-

———

[*] This is the DNA contained in the small bodies within a cell called *mitochondria*, which are responsible for converting glucose into available energy.
[†] This is the DNA contained in plastids, which are small bodies within plant cells where photosynthesis can take place. Plastids contain small circular chromosomes (of about a hundred thousand nucleotides) and are found outside the nucleus of the cell.

sult, the birds diversified into the 13 species now found on the islands.

Each species is adapted to its own niche. The shape and strength of their bills, as well as the muscles attached to them, are suited to the type of food they eat. The cactus finch, the woodpecker finch, the cocos finch and the warbler finch all have long and pointed bills. These birds probe flowers or foliage for food. Their bills are also good for spearing insects. The ground finch and the cactus finch have bills that are deep at the base. Their bills are good for crushing hard seeds and other hard food [Grant 1986].

How did the diversity of finches arise on the Galápagos Islands? Why do these finches differ from all other finches? Most experts agree with Darwin's assessment as modified by the NDT. They hold that a few finches once arrived at one of the Galápagos Islands from the mainland. The birds bred there and diversified through random variation and natural selection, filling many of the empty niches on these barren islands. The hypothesis is based on the assumption that small random mutations could have led to those adaptive changes. Is that assumption reasonable? Most evolutionists think it is, but some scientists don't think so. Ho and Saunders have said:

> It stretches credulity to imagine, for example, that the woodpecker first got a long beak from some random mutation followed by other random mutations that made it go in search of grubs in the bark of trees.

Moreover, it contradicts the calculations in Chapter 4, and there's no evidence for it. But if it couldn't happen the way the NDT says it did, then how could it have happened?

It could have come from nonrandom variation. It could have come through the direct influence of the environment. The type of food the birds ate as youngsters would determine their adult bill shape and jaw structure. Alternatively, it could have happened through a built-in genetic switch triggered by the environment. But all the evidence is against its happening through the process described by the NDT.

Laysan Island is a small coral island in the Pacific Ocean about a thousand miles northwest of Honolulu. It stands not much more than 10 feet out of the water. Laysan Island together with three other small islands in its vicinity form an official US Government Bird Reservation, which is the largest protected bird colony in the world. In 1967 about a hundred finches from Laysan were brought to a small atoll, called Southeast Island, about three hundred miles northwest of Laysan and about a hundred miles southeast of Midway Island. Southeast Island belongs to a group of four small islands all within a radius of about ten miles.

Through natural dispersion, and with some human help, the finches spread to all four islands of the group. When the birds were checked in 1984 they were already found to be different from the Laysan finches [Conant 1988, Pimm 1988]. By 1987 the population of finches had grown to about 800. When the birds were first put on Southeast Island in 1967 they were all alike. But when they were studied twenty years later, birds on different islands were found to differ from each other. In particular, they were found to have different bill shapes. The bills on the birds of North Island, about 10 miles north of Southeast Island, are deeper and shorter than those on either Southeast or Laysan. The birds on Southeast have longer bills than those on Laysan.

How did these differences arise so fast? To suggest the changes came from *random* variation and natural selection is unreasonable. The required mutations and the required selection could not have occurred in that short time. One might want to say that the necessary variation was already in the population when the 100 birds were introduced to Southeast Island. But if the variation was already in the population, why haven't those specimens been seen on Laysan Island in all the years the birds have been observed there?

The diversity of the finches on Southeast and North Islands

could, however, be explained as a phenotypic phenomenon entirely. A change of diet might produce the appropriate bill shape and jaw structure. Alternatively, the environmental effect could have thrown a genetic switch, which would have, in turn, changed the phenotype. In either case, the effect would appear in many individuals. There would be no waiting for a mutation to occur or for natural selection to work. A change to a new species could occur quickly, even in one generation.

Thomas Smith, of the University of California at Berkeley, studied an African finch in Cameroon. He found that the finches can produce offspring having two different bill sizes. He found this to be true in all three species of finch that he studied [Smith 1987]. He did not say whether or not the environment played any role in producing the two bill sizes. But he did report that the bill sizes are adaptive, each in its own niche. The birds with large bills crack large hard seeds easily, while those with small bills do so only with difficulty. The birds with small bills, on the other hand, feed more efficiently on small soft seeds than do those with large bills. These results show that bill size in finches can change from one adaptive type to another with diet.

In general, the more of its resources an animal puts into reproduction, the less its chance of personal survival. There is a theory that says if animals have a higher death rate as adults, then those that mature earlier and use a larger part of their resources on reproduction are more adaptive. If they have a higher death rate as juveniles, then the opposite is true: the more adaptive ones are those that mature later and use a smaller part of their resources for reproduction.

David Reznick and his team at the University of California have studied guppies in the wild. They found that guppies indeed behave as the theory predicts [Reznick et al. 1990, Reznick and Bryga 1987]. In a river with a predator fish that preys on large mature guppies, the guppies tend to mature early and have many small offspring. Where there is a predator fish preying on the small ones, the guppies tend to mature late and have fewer offspring.

There is a cichlid fish, of species *alta*, that preys on large mature guppies. The killfish preys on small immature guppies. Normally, guppies that live with the *alta* mature earlier and produce more and smaller offspring than do those that live with the killfish. The Aripo River in Trinidad has guppies together with *alta* cichlids, and the guppies that live there follow this rule. Reznick and his team took 200 guppies from the Aripo and put them in a tributary of that river that is home to the killfish but has no cichlids and had no guppies. Changes soon appeared in the newly introduced guppies. The fish population soon changed to what would normally be found in the presence of the killfish, and Reznick found the changes to be heritable.

The full change in the guppy population was observed as soon as the first samples were drawn, which was after only *two years*. One trait studied, the age of males at maturity, achieved its terminal value in only four years. The evolutionary rate calculated from this observation is some ten million times the rate of evolution induced from observations of the fossil record [Reznick et al. 1997].

Reznick interpreted these changes as the result of natural selection acting on variation already in the population. Could natural selection have acted so fast as to change the entire population in only two years? A more reasonable explanation is that the presence of the predator induced the changes in the individual fish.

Let me call the hypothesis suggested in this Chapter the *nonrandom evolutionary hypothesis*, and for brevity, let me call it just *NREH*. We have seen in Chapter 4 that the NDT does not account for the many so-called convergences in living things. What seem to be convergences can, however, be understood according to the NREH as the result of the action of the environment directly influencing the phenotype or the genotype. According to the NREH, what have been called convergences may not be that at all. They may be just similar built-in responses of different species to the same environmental cues. According to the NREH there is no surprise in finding two groups of organisms, in distant classes, having similar adaptive traits in similar environments. Seventy years ago Göte Turesson [1925] wrote,

> It has been known for quite a long time that plants distant in affinity but inhabiting the same habitat may also show one and the same, or similar, growth-forms.

The fish and the cephalopods are in different phyla. Yet when they occupy the same niche in the environment, they have many traits in common. These traits include their visual system, illumination systems, and locomotive systems. Fish in different environments differ among themselves in these traits, as do the cephalopods among themselves. But when the two phyla share a niche, they are also found to share common traits [Packard 1972].

There are many examples where species share common plastic traits. What appears to be convergence may just be the plastic response of the organism to its environment. Examples include the following.

- Limbs that protrude from an animal's body have more surface area per unit mass than the rest of the body. In cold weather the animal loses more heat per unit mass from these limbs than from other parts of the body. In many species the tails and legs are shorter for those living in colder climates and larger for those in warmer climates. Gulls' wings are shorter in cold climates than in warm. Hares and foxes also have shorter ears in cold climates than in warm. Eskimos have shorter arms and legs than do people living in warmer climates [Collier et al. 1973]. Sumner [1909] found that mice reared at low temperature had shorter legs and tails than mice reared at higher temperatures [Johnston and Gottlieb 1990].

- *Gloger's rule*: Races of birds or mammals living in cool dry regions have lighter skins than do races of the same species living in a warm humid area [Schreider 1964]. This is true of humans as well.

- *Jodan's rule*: Many species of fish tend to have more vertebrae when they live in cold water than do the same species living in warm water [Schreider 1964]. These differences have been shown to depend on the temperature at which the fish have been reared [Brooks 1957; Hubbs 1922, 1926; Murray and Beacham 1989, Johnston and Gottlieb 1990].

What these rules show is not *convergence*. They show that different species adopt the same anatomical strategies when they have to cope with the same environmental conditions. We have seen that these strategies cannot come from random mutations. It is much more reasonable to say they come from environmental cues acting on the genetic program. The hypothesis seems particularly reasonable when we observe these effects in many plants and animals.

The NREH is a hypothesis that explains many observed phenomena that the NDT does not explain. According to the NREH, adaptive modifications in organisms occur when the environment induces a change in either the phenotype or the genotype. It can account for the environmentally induced adaptive mutations reported in bacteria. It can account for the pervasive convergences found throughout the plant and animal kingdoms. The NREH does not suffer from the contradictions of the NDT, and promises therefore to provide a more consistent picture of life.

Chapter 8

EPILOGUE

THE forgoing chapters have established two major points about evolution. The first several chapters have shown that random variations cannot lead to the large-scale evolution claimed by the neo-Darwinians. The seventh chapter has shown that there is a lot of evidence for *nonrandom* variation, which could produce some large-scale evolution. I offered there what I called the *nonrandom evolutionary hypothesis (NREH)* to account for this evidence.

Randomness is an essential feature of the NDT. There is no known physical or chemical mechanism to generate heritable variations that will improve adaptivity or increase the complexity of living organisms. The neo-Darwinians, therefore, had to choose randomness to produce the variations they needed. In this way they hoped that, through the direction afforded by natural selection, they could describe an evolutionary process that could account for a natural origin and development of life.

The neo-Darwinians have rejected *non*randomness as the major feature of variation. They departed from Darwin on this point. Although, at first, Darwin chose chance as the only determining factor for the variation, he later rejected that position.

Darwin was apprised that it is likely for even the most favorable adaptation to die out before it can take over the population by natural selection [Himmelfarb 1962, p. 321].[*]

---

[*] Note that this was shown mathematically by Fisher [1958] as discussed in Chapter 3.

209

He responded by distancing himself from randomness. He said the variations were not just chance, but there is a built-in tendency for them to vary under the influence of the environment. Darwin reneged on attributing variation to chance, and said,

> I have hitherto sometimes spoken as if the variations ... were due to chance. This, of course, is a wholly incorrect expression, but it serves to acknowledge plainly our ignorance of the cause of each particular variation. ... variability is generally related to the conditions of life to which each species has been exposed during several successive generations. [Darwin 1872, p. 128]

Doesn't this look like the NREH?

Without randomness, the NDT has no mechanism to offer for evolution. Natural selection alone is powerless without the raw material of variation. As Samuel Butler has said, "The 'Origin of Variation,' whatever it is, is the only true 'Origin of Species.'"[*] Random variations cannot lead to the kind of evolution the neo-Darwinians sought.

The NDT insists that the origin of all life, including our own, is grounded in randomness. The NREH, on the other hand, postulates nonrandom variation. It does not fulfill the neo-Darwinian agenda in that it does not contribute to a natural explanation of the origin of life. But then, neither does the NDT. Like it or not, we don't have a theory that can account for a natural origin of life.

If we ignore the neo-Darwinian agenda, we see that the NREH can account for observations of evolution better than can the NDT. It can account for observations that the NDT fails to account for. It can account for the environmentally directed mutations reported in bacteria. It can account for the numerous so-called convergences. If we simply drop what has been the requirement of the NDT that evolutionary theory justify a natural origin of life, then the difficulties of evolutionary theory fall away, and we can arrive at a better under-

---

[*] Butler [1878], as quoted by Himmelfarb [1962]

standing of the life sciences. We can begin to make better sense of the vast amount of biological data that is pouring out from laboratories around the world.

One cannot fault the NREH because it fails to account for the development of life from a single cell. Such a development has not been observed, so there is no imperative for a theory to account for it. The NREH accounts for what has been observed, and it does so better than the NDT. Those are adequate credentials for a theory of evolution.

Had the NDT been successful, together with a theory to account for a natural origin of a simple cell, we would have had a good natural explanation of life. If so, a supernatural explanation would have been superfluous. The NDT would then have put Jewish tradition and the Torah view of the origin of life on the defensive.

The defense would then have had two options. The *first option* would be just to reject the scientific view as incorrect, simply because it arrives at the wrong conclusion. Although this option may seem low-brow, it is robust and has proved its value in the past.

For example, until not much more than 50 years ago, the accepted scientific opinion was that the universe was infinitely old and had no beginning. This scientific view dated from Aristotle and denied the Torah concept of creation. Torah scholars unanimously rejected the infinite-age theory, in spite of all the scientific authority behind it. In rejecting it, they have adopted this first option.

The past 50 years have seen in science the replacement of the infinite-age theory with the big-bang theory of the origin of the universe. The basic position of the Torah scholars has thus been vindicated by recent scientific discoveries, even though there is still disagreement about time scale.

The *second option* would have been apologetic. It would accept the theory, and try to fit the Creator into the randomness. What the theory calls random, it would attribute to the will of the Creator. It would do exactly what the Scottish clergyman Henry Drummond a century ago ridiculed the apologists for. He mocked those who "ceaselessly scan the fields of Nature and the books of Science in search of gaps, gaps which they will fill up with G-d." [Drummond 1894, p. 426; Himmelfarb 1962, p. 393].

Had the Torah scholars adopted an apologetic attitude toward the scientific theory of an infinitely old universe, they would have been forced to make an embarrassing reversal during the past 50 years.

Although there are defensive positions the religious believer could have taken against the NDT, the theory *does* deny creation. The NREH, on the other hand, is agnostic and poses no contradiction to creation. The NREH, as an explanation of evolution, is in fact derivable from Talmudic sources.

Rabbi David Luria (1798-1855), known as the רד״ל, indeed made such a derivation in his commentary to the Midrash פרקי דרבי אליעזר [Luria 1990, pp. 52a-b]. From Talmudic and Midrashic sources he derived the necessity of animals to evolve. As Rabbi Luria interpreted the Midrash, there were 365 basic species (מינים) of beasts created, and the same number of birds. All the others were derived from these. As each basic species moved into a different environment and found itself a new niche, it changed. The changes were dictated by the conditions under which it lived, including the food it ate. Rabbi Luria's conclusion is very much like the NREH presented in Chapter 7.

We have seen, then, that the NDT, which describes an evolution that contradicts the Torah's concept of creation, is incompatible with many scientific observations. The NDT not only stands in the way of a better understanding of the life sciences, it also tends to prevent us from appreciating that there may be spiritual values in the universe that stem from a source higher than man. The NREH, on the other hand, can embrace the Torah's view of creation and evolution, and accounts well for scientific observations.

תושלב״ע

Appendix

A. THE STRUCTURE OF DNA

THE DNA (deoxyribonucleic-acid) molecule is a polymer (a chain) whose elements (the links) are *nucleotides*. A *nucleotide* is made up of what is called a *nucleoside*, which is the main part of the link, and a phosphate group, which serves as the connector joining the links to each other. The phosphate group consists of a phosphorus atom joined with four oxygen atoms. The *nucleoside* is a combination of two molecules. One is a sugar, the other is called a *base*.

In DNA the sugar is *deoxyribose* (that's where DNA gets its name). The deoxyribose sugar is a molecule of 19 atoms of carbon, oxygen, and hydrogen. The bases are molecules built from 12 to 16 atoms of carbon, oxygen, hydrogen, and nitrogen.

There are four kinds of bases, and therefore four kinds of nucleotides. The four bases are: *adenine*, which we denote by A, *thymine*, denoted by T, *cytosine*, denoted by C, and *guanine*, denoted by G. The nucleotides are strung together in a long chain with a mixture of the four kinds of bases. The DNA can carry information in the order of its bases.

Fig. A.1 is a schematic diagram of the DNA molecule. The molecule consists of two parallel strands of the polymer connected to each other. The sugars (the circles labeled "SGR" in the figure) are linked by phosphate groups (the small circles) to form the *backbone* of the strand. The bases are shown as larger circles and are labeled A, C, T, or G. Each base is attached to a sugar molecule and comes off at right angles to the backbone. The bases of the two strands extend toward each other and are joined through a chemical bond called a *hydrogen bond*. The hydrogen bonds are shown in Fig. A.1 as dashed lines. A hydrogen bond is formed between two atoms when they share an atom of hydrogen. These bonds serve to hold the two strands together.

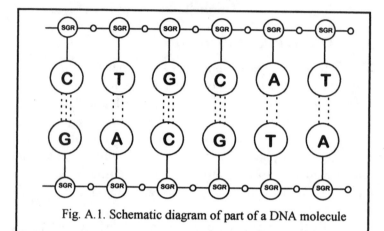

Fig. A.1. Schematic diagram of part of a DNA molecule

In the figure, the backbones of the two strands go off the page both to the left and to the right. One of these strands runs across the top of the figure, and the other runs across the bottom. The two strands are twisted into a helical shape to form the famous double helix.

Think of the figure as extending to the left and right many millions of times the length shown. The DNA molecule is much like a long narrow ribbon, or rather a pair of ribbons, one for each backbone. The bases are connected to the backbone at fixed intervals along the length of the molecule.

The bases on the two strands of a DNA molecule are restricted in the way they pair with each other. The size and shape of the bases and the way they form their hydrogen bonds determine which base will pair with which. An A on one strand will pair only with a T on the other, and a C will pair only with a G.

The two strands of DNA are related in this way all along their length. Because of this restricted pairing, the bases of the two strands are said to be *complementary* to each other.

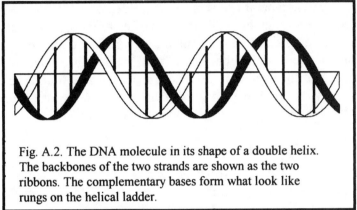

Fig. A.2. The DNA molecule in its shape of a double helix. The backbones of the two strands are shown as the two ribbons. The complementary bases form what look like rungs on the helical ladder.

Fig. A.2 shows the DNA molecule on a slightly smaller scale than that shown in Fig. A.1. The two helical ribbons run across the page with the base pairs connecting the two strands like the rungs of a ladder. Indeed, the DNA is much like a helical staircase.

B.   STORAGE OF INFORMATION

A binary digit, or symbol, is one that can take on either of two values, usually denoted by 0 (ZERO) and 1 (ONE). The amount of information that can be carried by one binary digit is defined as one *bit.* The volatile memory of a computer uses 5 volts to represent ONE, and 0 volts to represent ZERO. A memory chip that holds eight binary digits, for example, can hold eight bits of information. (In computer jargon, a group of eight bits is called a *byte*.)

A symbol of a large alphabet can hold more information than can a symbol of a small alphabet. For example, a letter of the English alphabet can hold up to 4.7 bits of informa-

---

* John Tukey, a mathematician at Princeton University coined this term about 50 years ago. He chose the term *bit* as a contraction of *binary digit*.

tion. A symbol of a binary alphabet, on the other hand, can hold up to 1 bit.* For a device to store a symbol of a 26-letter alphabet, it has to have 26 stable states.† To store a symbol of a *binary* alphabet it needs only *two* stable states. A DNA nucleotide has four states one for each of its four possible bases. So we could call a DNA symbol a *quadrinary* digit, and we can encode it with 2 binary digits. (I made up this word to mean that the symbol is from an alphabet of size four. Alternatively we might call it a *quadrit*, or maybe a *quit*.) Each DNA symbol can hold up to 2 bits of information.

C. HOW ENZYMES WORK

How do enzymes work? How does an enzyme make a chemical reaction go much faster than it would without the enzyme?

A protein's shape is not always fixed. It can change as it is functioning. Sometimes a small molecule attaches to the protein and causes its shape to change. The change in shape often causes the protein's behavior to change as well. In fact, the cell often controls the protein's behavior through such small molecules. You can think of an enzyme as a tiny molecular machine. It grabs the molecule it's supposed to work on (its *substrate*) and distorts it. In the language of the physical chemist, it stabilizes the transition state. It thereby speeds up the reaction by a large factor. Let us see what this means.

The science of thermodynamics tells us that a reaction tends to proceed in the direction in which the system drops from a state of higher energy to a state of lower energy. As it does, it releases the difference in energy between the initial

---

* The English alphabet has 26 symbols. A binary alphabet has two symbols. A symbol of the English alphabet can hold about 4.7 bits because $2^{4.7}$ is about 26. A symbol of a binary alphabet can hold 1 bit, reflecting the fact that 2^1 is 2.

† If we count the space as well as the 26 letters of the alphabet, there are 27 symbols. A device that can store any letter or a space must have 27 stable states

and the final state. This would seem to say that glucose ought to be unstable. So if I put glucose in a dish, why doesn't it automatically burst into flame, burn down to CO_2 and water, releasing the energy of the reaction in the form of heat?

As an example, let's look at the metabolism of glucose. Glucose is a six-carbon sugar and is the main source of energy in cells. The glucose molecule, like all sugars, is made of carbon, hydrogen, and oxygen, which are joined by chemical bonds. These bonds have potential energy in them, and it is this energy the cell seeks when it feeds on the sugar. The cell oxidizes glucose, breaking the chemical bonds in it, and producing carbon dioxide (CO_2) and water (H_2O). The energy released in breaking those bonds is called the *reaction energy*.

Glucose doesn't spontaneously burst into flame because it can't convert all its chemical bonds at once into the bonds of the product, CO_2 and water. For a substrate to change to the product, it often has to pass first through a *transition state* whose energy is higher than that of its initial state. The energy of the substrate has first to rise to that of the transition state. Only then can the reaction drop to the product state.

Energy must be added for the substrate to rise to the transition state. If we should add that extra energy, called the *activation energy*, to the glucose, say by igniting it with a match, it will rise to the transition state and burn. It will give off heat as it falls from the transition state of high energy to the lower one of the products.

The system energy starts at the level of the initial state as shown in Fig. A.3. It can't drop to the level of the product state because there's a barrier in the way. The height of the barrier is the energy of the transition state. If we want the system to drop to the energy of the product level, we first have to add enough energy to bring it up to the transition state. Then it will descend of its own accord. That's why we have to ignite the glucose to get it to burn.

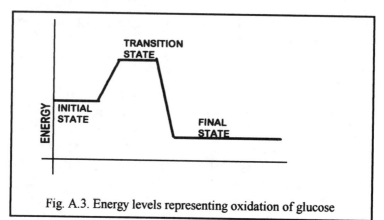

Fig. A.3. Energy levels representing oxidation of glucose

Random knocks from other molecules (thermal energy) can once in a while give a substrate molecule enough energy to raise it up to the transition state. But knocks of large energy are rare. The reaction will indeed go by itself, triggered by random thermal energy, but it will take a long time if the activation energy (the height of the barrier) is large. The reaction time is a sensitive function of the height of the barrier. If the barrier were one energy unit higher, the reaction time would be a factor of e (about 2.7) larger. If it were one unit lower, the reaction time would be less by the same factor.[*]

The role of the enzyme is to lower the energy level of the transition state. The dashed line in Fig. A.4 shows what the transition energy level might look like with the enzyme. If the enzyme lowers the energy of the transition state by 14 units, the reaction rate will increase by e^{14}, or a factor of a million. If it lowers the energy by 32 units, the reaction rate will increase by a factor of e^{32}, a factor of a hundred trillion.

---

[*] I've taken here as the unit of energy per molecule its mean thermal energy, which is kT, where k is Boltzman's constant and T is the absolute temperature [Savageau 1976].

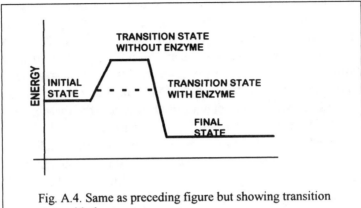

Fig. A.4. Same as preceding figure but showing transition state with the enzyme

The enzyme reduces the energy level of the transition state by binding to the substrate and changing its shape [Stryer 1988, p. 184]. Reducing the activation energy speeds up the reaction. The enzyme's shape is specific to the substrate. It fits the substrate, and does not fit other molecules. The specificity keeps the enzyme from taking up and working on competing molecules.

A reaction is usually carried out in more than one step, with a separate enzyme for each step. Each step changes the substrate to become the input for the action of the next enzyme. When glucose is oxidized in the cell, it goes through many steps, mediated by about 30 different enzymes. Energy is taken off these reactions in small doses and each is packaged into a small energetic molecule of ATP (adenosine triphosphate). Thirty eight such packages are formed during the conversion of each glucose molecule. Two are used in the process, so the burning of one molecule of glucose gives a net yield of thirty six of these small energy packages.

Each step is mediated by an enzyme that reduces the activation energy of the transition state enough to permit the process to be carried out at body temperature. Contrast this with the high temperature at which the glucose burns when you set a match to it.

Once the enzyme binds to the substrate, it applies forces to it, distorting it into a low-energy transition state. It also adjusts its own shape to bind even more closely to this transition state of the substrate. From the transition state, the substrate quickly passes to the product state and the reaction is over. The products take leave of the enzyme, which returns to its resting state. The enzyme is then ready for the next substrate molecule. The whole process can take a ten thousandth of a second. The ability of proteins to bind specifically to almost any biological molecule, and lower its transition-state energy, makes enzymes powerful machincs that have a far-reaching influence on all living processes.

D. HOW THE CELL CONTROLS CHEMICAL REACTIONS

Biochemical reactions often convert an initial substrate S_0 into a final product in several steps. Each step is catalyzed by

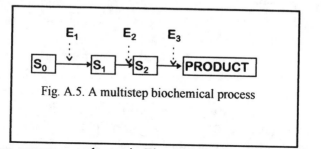

Fig. A.5. A multistep biochemical process

its own enzyme, as shown in Fig. A.5. In this figure, enzyme E_1 catalyzes the conversion of S_0 into S_1, and so on to the final product.

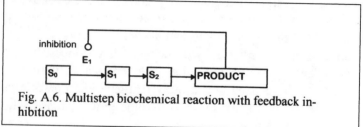

Fig. A.6. Multistep biochemical reaction with feedback inhibition

The final product then usually inhibits the enzyme E_1 that catalyzes the first step, as shown schematically in Fig. A.6. The inhibition caused by the final product in the figure going back and inhibiting an earlier enzyme is called *feedback inhibition*. The small circles in this and subsequent figures indicate inhibition. This first step in a multistep biochemical reaction is called the *committed* step.

Sometimes the path of a biochemical reaction is branched, as shown in Fig. A.7. The initial substrate S_0 leads to two products P_1 and P_2. If P_1 were to inhibit the formation of S_1, then an abundance of P_1 would cut off the production of P_2 as well as P_1. To avoid this pitfall, P_1 inhibits only the production of S_3, so that an abundance of P_1 does not cut off production of P_2. The product P_2 inhibits the formation of S_5. An accumulation of S_2 inhibits the formation of S_1. This arrangement prevents any excess buildup of intermediaries.

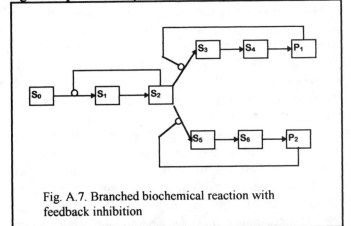

Fig. A.7. Branched biochemical reaction with feedback inhibition

Another method sometimes used in biochemical systems is that shown in Fig. A.8. Here there are two ways of performing the committed step, each catalyzed by its own enzyme. The product P_1 inhibits one pathway to S_1, and P_2 inhibits the other. This arrangement prevents a buildup of S_1 only when there is an abundance of both P_1 and P_2. The product P_1 also

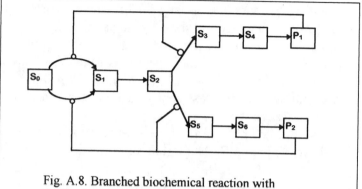

Fig. A.8. Branched biochemical reaction with
feedback inhibition different from that in Fig. A7

inhibits the production of S_3, and P_2 inhibits the production
of S_5.

In these examples, the product molecule controls the activity of the enzyme. This control is often effected through what
is known as *allosteric*[*] interaction. The control molecule, in
binding to the enzyme, changes the enzyme's *shape*. The
change in shape shuts OFF the enzyme activity. Thus as
product builds up, the enzyme activity gets shut OFF and no
more product is made. The control is reversible, so that when
the product gets used up, the enzyme activity returns [Stryer
1988, pp. 239-ff.]. All these examples of feedback inhibition
are found in living cells. Control of enzyme activity contributes to the efficiency of the cell's biochemical functions.

Some enzymes are controlled by special proteins. The control can be either positive (stimulatory) or negative
(inhibitory). An example of such a protein is *calmodulin*,
which regulates many enzymes. Calmodulin itself is turned
ON by the presence of calcium ions. When calmodulin is active it turns ON a variety of other enzymes.

Other control proteins keep enzymes OFF. The enzymes under control all have special sites on them to which the control

---

[*]*Allostery* is a word coined especially to apply to the action of these
enzymes. The word is derived from the Greek roots *allos*, meaning
"other" and *stereos*, meaning "solid".

proteins can bind. Still other enzymes are activated when a piece of them is chopped off to expose an active site. Digestive enzymes are activated in this way, as are enzymes that aid in blood clotting.

E.   TRANSCRIPTION OF DNA TO RNA

I shall first describe how transcription is done in bacteria, and then I shall note some additional features found in other organisms. As the first step in the making of a protein, the cell builds an mRNA polymer as a complementary copy of the sequence of nucleotides in the DNA.

Building the mRNA is a job of copying information, and you can think of it as something like copying a magnetic tape. An enzyme called *RNA polymerase* catalyzes this job [Darnell et al. 1986, pp. 124 ff.]. As far as we know, there is an enzyme to catalyze every chemical function of the cell. The cell uses a special enzyme to control each step of every molecular process it performs.

The polymerase searches along the DNA molecule until it finds the gene it wants to copy. It looks for a special base sequence called the *promoter*. The promoter acts like a flag that signals the start of the gene. It is also where the polymerase first attaches. Fig. A.9 shows schematically the polymerase sitting on the promoter. The promoter is near, but not on, the section of the DNA sequence to be copied.

The polymerase starts at the promoter, unwinds, and opens a section of the double helix of the DNA as shown in Fig. A.9. To open the double strand, it has to break the hydrogen bonds that join the complementary bases. When it breaks these bonds, it exposes the bases of the individual DNA strands, permitting one of them to be copied.

The copying is done by pairing RNA bases with their complements on one of the DNA strands. The RNA nucleotides are brought to the polymerase for it to pair them with those of the DNA.

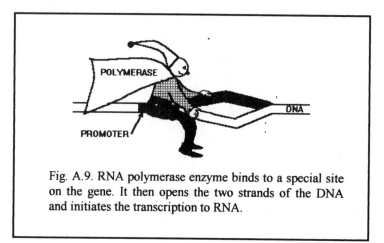

Fig. A.9. RNA polymerase enzyme binds to a special site on the gene. It then opens the two strands of the DNA and initiates the transcription to RNA.

The polymerase moves along the DNA, and as it moves it unwinds more of the DNA helix and separates the two strands. As more RNA nucleotides are brought into position, the polymerase lines them up by pairing them with the DNA nucleotides in the gene. It then connects these nucleotides to each other, building a ribbon of mRNA as it moves forward. As it moves, it opens a new section of the DNA and closes the old section, letting the helix rewind behind it. Usually only one strand of the DNA is copied. But sometimes both are [Normark et al. 1983].

As the mRNA molecule grows, its finished end peels off the DNA. The string of nucleotides come off the DNA soon after they attach to the mRNA chain. At any instant there are only about a dozen mRNA bases attached to their complements on the DNA. Even before the mRNA is completed, its finished end joins with a ribosome and starts to direct protein synthesis. It's as if the new-born mRNA molecule can't wait to get to work.

The polymerase continues to copy the gene until it comes to a *stop code*. When it gets to the stop code, it stops copying and releases the rest of the finished mRNA ribbon from the DNA.

So far I have told you how bacteria copy the information from the DNA to the mRNA. Bacterial cells do not have nuclei and are called *prokaryotes*. All other kinds of cells do have nuclei and are called *eukaryotes*. The eukaryotes code their information in a more complex way than do the bacteria. So the job of handling the information in eukaryotes has to be more complex than it is in bacteria. There are two main differences between the genes of eukaryotes and those of prokaryotes:

- First, eukaryote genes are in the nucleus of the cell. The membrane of the nucleus tends to isolate its contents from the rest of the cell.

- Second a eukaryote gene does not encode a protein in a continuous mapping as does a prokaryote gene. In the eukaryote, the DNA encoding the protein has interspersed within it sections of DNA whose purpose we do not yet know.

In eukaryotes the segments of the DNA that encode protein are known as *exons* because they're *ex*pressed. They're expressed in the proteins they encode. The segments between the exons are known as *introns*. Walter Gilbert of Harvard University gave them this name because they are *intra*genic: they lie "between the genes" [Gilbert 1978]. Many eukaryote genes are split up in this way. Most genes of the higher eukaryotes, such as birds and mammals, are split. Fewer genes of the lower eukaryotes, such as yeast and fungi, are split. The relative lengths of exons to introns is variable. In some cases the total length of the introns can be twenty times the length of the exons.

Eukaryotes and bacteria differ somewhat in how they copy information from the DNA to the protein. Eukaryotes make their mRNA in stages. In the first stage, the cell copies the entire gene, both the exons and the introns, onto the mRNA. In the second stage the cell cuts out and removes the introns from the mRNA. Then it splices the exon pieces together to close the gaps. The mRNA, with only the exons in it, leaves the nucleus and joins with a ribosome to make protein [Darnell et al. 1986].

F. THE GENETIC CODE

The DNA contains the information that tells the cell how to make its proteins. It carries a coded list of the string of amino acids making up each protein the cell may have to assemble. The amino acids are encoded in the DNA in the same order in which they appear in the protein. The DNA encodes each amino acid with a string of three bases, and the RNA does the same. A triplet of bases that encodes an amino acid is known as a *codon*. A codon encodes one amino acid. A molecule of mRNA will encode a complete protein. A typical protein of 300 amino acids is encoded by an mRNA having 300 codons, or 900 nucleotides. This code can be traced back to 900 nucleotides in the DNA.

There are 64 different codons corresponding to the 64 ways of forming a triplet of DNA symbols[*]. These 64 codons encode the 20 amino acids that make up the proteins. Note that there are more codons (64 of them) than there are amino acids (20 of them). More than one codon can therefore encode the same amino acid. For any code, if the same clear text can be encoded in more than one way, the code is called *redundant*. The genetic code is redundant.

Table A.1 shows the genetic code. It shows the translation of each RNA codon into its corresponding amino acid. The abbreviations for the amino acids are the standard ones. Note that in the table there are amino acids for only 61 of the codons. Three are translated to STOP. These codons do not code for an amino acid. Instead, they play the role of *stop codes*, which mark the end of a protein code.

---

[*]There are 4 DNA symbols. There are therefore 4×4×4, or 64, ways of forming a triplet.

The bases play the role of the *letters* of the DNA alphabet, and the genes play the role of the DNA *words*. In the *protein* alphabet, the *amino acids* play the role of the *letters*, while the proteins themselves play the role of *words*. Three DNA letters translate into one *protein* letter, and one DNA word encodes one protein word. The stop codes in the DNA translate into the *spaces* between the protein words.

UUU	Phe	CUU	Leu	AUU	Ile	GUU	Val
UUC	Phe	CUC	Leu	AUC	Ile	GUC	Val
UUA	Leu	CUA	Leu	AUA	Ile	GUA	Val
UUG	Leu	CUG	Leu	AUG	Met	GUG	Val
UCU	Ser	CCU	Pro	ACU	Thr	GCU	Ala
UCC	Ser	CCC	Pro	ACC	Thr	GCC	Ala
UCA	Ser	CCA	Pro	ACA	Thr	GCA	Ala
UCG	Ser	CCG	Pro	ACG	Thr	GCG	Ala
UAU	Tyr	CAU	His	AAU	Asn	GAU	Asp
UAC	Tyr	CAC	His	AAC	Asn	GAC	Asp
UAA	STOP	CAA	Gln	AAA	Lys	GAA	Glu
UAG	STOP	CAG	Gln	AAG	Lys	GAG	Glu
UGU	Cys	CGU	Arg	AGU	Ser	GGU	Gly
UGC	Cys	CGC	Arg	AGC	Ser	GGC	Gly
UGA	STOP	CGA	Arg	AGA	Arg	GGA	Gly
UGG	Trp	CGG	Arg	AGG	Arg	GGG	Gly

Table A.1. The genetic code giving the amino acid corresponding to each RNA triplet.

G.   HOW PROTEINS ARE MADE

A ribosome is a body within the cell that uses the information in the mRNA to make protein. The ribosome is made of a combination of protein and RNA. It reads the information in the mRNA, a codon at a time, and joins amino acids together in a chain as dictated by the codons. More than one ribosome can work on the same mRNA molecule at the same time, each making its own protein.

Special small RNA molecules, called *transfer RNA (tRNA)*, bring the amino acids to the ribosome. There are many different tRNA's, nearly one for each codon. Some codons that encode the same amino acid share a tRNA. (Recall that there are 61 codons that encode 20 amino acids.)

Each tRNA carries only its own amino acid. It has a site that matches to its amino acid, and another that matches to the mRNA codon. The latter site is a codon whose three nucleotides are complementary to the mRNA. Some tRNA's belong to more than one codon, and they may match only the first two of the three codon bases.

The ribosome puts the amino acids together to make a protein. A tRNA attaches to an amino acid with the help of a special enzyme. It then brings the amino acid to the ribosome that's making the protein.

Fig. A.10 shows the ribosome in the midst of building a protein. Part *a* of the figure shows the ribosome adding the amino acid serine (Ser) to the protein it's building. Only the tail end of the protein appears in the figure. The rest, which goes off the figure to the left, is not shown. Note the tRNA holding the serine on its bottom. On its top is the codon UCG. This codon binds to the complementary codon AGC on the mRNA. (From Table A.1 you can see that AGC encodes serine.) The tRNA is then in the right position to deliver its amino acid. The figure shows the tRNA delivering serine and adding it to the protein. The ribosome joins the serine to the growing protein through a peptide bond.

228

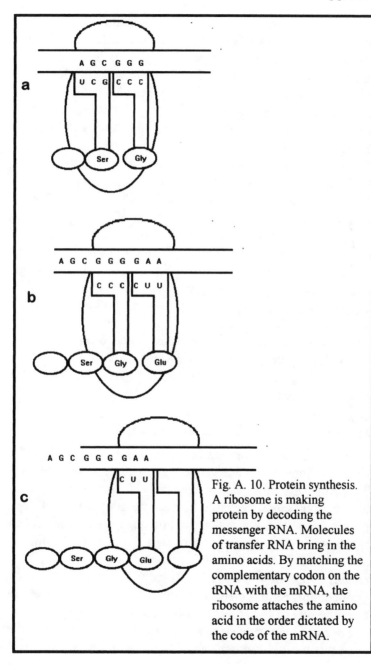

Fig. A. 10. Protein synthesis. A ribosome is making protein by decoding the messenger RNA. Molecules of transfer RNA bring in the amino acids. By matching the complementary codon on the tRNA with the mRNA, the ribosome attaches the amino acid in the order dictated by the code of the mRNA.

While the ribosome is connecting the serine to the protein, another tRNA is busy bringing in the next amino acid. In the figure, this tRNA carries glycine (Gly) on its bottom, and on its top it has the codon CCC to match the codon GGG on the mRNA. (From Table A.1 you can see that GGG encodes glycine.) Part *b* of Fig. A.10 shows the ribosome attaching glycine to the protein. Meanwhile, another tRNA is bringing in the next amino acid, glutamic acid. Part *c* of the figure shows the ribosome joining the glutamic acid to the protein.

The tail of the new amino acid forms a peptide bond with the head of the last one added. The tRNA then splits off from the mRNA and a new tRNA comes in. When the protein is complete, the ribosome disengages, and the new protein folds into its functional shape. Protein folding is helped by special small proteins known as *chaperones* [Craig 1993, Agard 1993]. These proteins had previously been grouped with a larger class of proteins called *heat-shock proteins*, or *stress proteins* [Welch 1993] until scientists discovered their special role as protein folders. Stress proteins had been discovered about thirty years ago to be the cell's defense against heat damage. Since then, they were discovered also to defend the cell from more general stress.

H.   HOW THE CELL REGULATES ITS GENES

Some gene regulators act in a positive way by turning the gene ON. Some act in a negative way by turning the gene OFF. Those that act in a positive way are called *activators*. They help the polymerase enzyme bind to the gene. In effect, they turn the gene ON. Those that act in a negative way are called *repressors*. They repress, or turn OFF, the gene by binding to a site on the DNA near the promoter. The binding site is called an *operator*. Fig. A.11 shows the repressor sitting on the operator. Sitting there, it denies the polymerase access to the promoter. The repressor will sit there as long as the cell doesn't need the gene's protein. While the repressor is on the operator, the polymerase can't attach to the promoter, so it

Fig. A.11. The repressor protein prevents the polymerase enzyme from binding to the promoter, preventing RNA synthesis.

can't start copying the DNA. The repressor, then, keeps the gene turned OFF.

When the cell needs the protein encoded by the gene, it turns the gene ON. When it wants to stop making the protein, it turns the gene OFF. For the cell to turn the gene ON it must remove the repressor, which it often does by means of a molecule known as an *inducer*. Fig. A.12 shows the inducer turning the gene ON by removing the repressor.

A regulatory protein, like all proteins, is encoded in a gene. A gene encoding such a protein is known as a *regulatory gene*. A regulatory gene sits in the genome near the gene that it regulates.

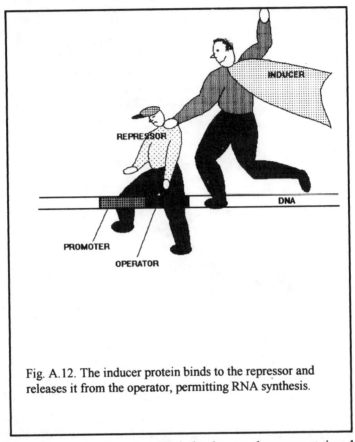

Fig. A.12. The inducer protein binds to the repressor and releases it from the operator, permitting RNA synthesis.

Sometimes small molecules help the regulatory proteins do their job. These are often just the encoded enzyme's *substrate* and *product*. The substrate often helps turn the gene ON, and the product often helps turn it OFF. The cell makes the enzyme only when there is substrate. It stops making it when there is either enough product or no more substrate. This control is another example of negative feedback, or *feedback inhibition*. The control ensures that the cell will make only as much enzyme as it needs and only when it needs it.

We have seen in Appendix D that the cell regulates the activity of its enzymes. Here we see that it also regulates their production. The cell is stimulated to produce and activate an enzyme when, and only when, it needs it. When the cell no longer needs the enzyme it stops making it and deactivates whatever enzyme is left. This two-pronged control makes the genetic system efficient and responsive to the cell's needs.

Most biochemical reactions in a cell are not executed in just one step. They are often carried out in a series of steps using a special enzyme for each step, as we have seen in Appendix D. The cell then uses a series of enzymes to run a reaction. The genes that encode such enzymes are often next to each other on the genome, sitting in the same order along the DNA as the order in which their enzymes act. The cell regulates these genes as a unit, with a common promoter for all of them. A series of genes under common control, like this, is called an *operon* [Darnell et al. 1986].

I.   *EMBRYOLOGICAL DEVELOPMENT*

Embryologists have long suspected that the genome contains a set of programmed instructions for developing the embryo. Molecular biologists are now finding the evidence for such a program, and they are discovering how it works.

As cells divide, the embryo grows and develops. Developmental biologists are learning how enzymes control cell division [Murray and Kirschner 1991]. They have found proteins that trigger the cell and cause it to divide during embryonic development [O'Farrell et al. 1989]. There are many of these triggers, and each acts at its own appointed stage. The zygote itself makes the proteins that trigger the first few cell divisions. The proteins that trigger the later divisions are made by the embryo. Although we don't yet know how they work, we do know that each trigger acts at the right time to make the embryo develop according to plan.

I am always amazed when I look at a new baby grandchild (I don't remember being so amazed when my own children were small). The perfection of the eyes, the limbs, even the finger nails and the eyelashes. How did this perfect and enormously complex organism develop from a single cell? What kind of a program orchestrates this development. Eyes that see, ears that hear, a brain that can think and command the movement of complex organs, have all developed from a single cell. That each of us started as a single cell and developed into a functioning, thinking, human being is mind-boggling.

How does the development program work, and how is it so perfect? As we have seen, development starts with one cell, which divides into two. The two divide into four, and so on. At some point, the cells differentiate, which means they change their character. Eventually, some of the cells become one kind of tissue or organ, while other cells become other kinds. And it happens the same way nearly every time. How does it happen?

The first question is how does the zygote, a single cell, give rise to different cells. If a single cell divides to yield two genetically identical daughter cells, then why doesn't the symmetry persist to keep all cells identical? Regardless of the nature of the development program, the cells can't differentiate unless the symmetry is somehow broken. How is that done?

It turns out that the symmetry is already broken in the zygote before it divides for the first time. When it divides into two cells, those cells are not perfectly identical.* Before the zygote divides for the first time, its cytoplasm is not uniform. Cells neighboring the zygote, called nurse cells, make proteins which they transfer into the zygote in a nonuniform pattern. Because the cytoplasm of the zygote is not homogeneous, the cytoplasm of its two daughter cells differ. These proteins eventually act on the genome to break the genetic symmetry.

---

*Even though the two cells are not identical, if they are separated, each will give rise to a complete individual. Identical twins are made in this way.

Some of these proteins have been discovered in *Drosophila*. The zygote has been found to have a gradient of protein density along the line that is to become the axis of the fly running from head to tail (the anterior/posterior, or A/P, axis). Two proteins, called *bcd* and *nos*, have gradients in opposite directions. A suitable developmental program can, with the aid of these gradients, and using positive feedback, produce arbitrary differentiation along the A/P axis of the embryo. There is a similar symmetry breaking along the back-to-belly axis (dorsal/ventral, or D/V, axis), but how it works is not yet known in much detail [Griffiths et al. 1993].

In the process of differentiation, the cell passes from one state to another according to the program in the DNA. In some cases, cells pass autonomously from one state to the next. The next state depends only on the current state. In other cases, the state to which a cell will pass depends on the signals it gets from the outside. Its next state depends both on its current state and on the states of its neighbors, which are signaled to it by messenger molecules. In some cases it depends also on the state of the environment.

The state of a cell is determined by the types and amounts of proteins it makes [Griffiths et al. 1993]. Through the proteins it makes, the cell can be characterized as a liver cell or a stomach cell, a skin cell or a muscle cell. The state of the cell pertains to the role it plays at any instant. In the adult, the state of a cell refers to its functional role. During development, the state refers to each condition the cell passes through on its way to developing into its ultimate role in the adult animal. Particularly important in determining the cell's state are the proteins that make up the developing embryo's system of information transmission. These are the proteins a cell makes that serve as regulators to its genes, that serve as outgoing signals to other cells, and that serve as receptors to detect incoming signals. The signals a cell receives lead to the turning ON or OFF of selected genes.

The differentiation process starts with the gradients of the *bcd* and *nos* proteins causing the expression of what are called *gap* genes in the early embryo.* The gap genes are responsible for determining the number of segments in the fly embryo. The gap genes in turn cause the expression of the *pair-rule* genes. These genes are responsible for making segment pairs. Then there are *segment-polarity* genes that set the A/P polarity of the segments.

The gap genes also cause the expression of what are called *homeotic* genes. The homeotic genes control the identity of the individual segments. The homeotic genes actually operate in complexes of several genes. In *Drosophila*, these genes have been discovered, identified, and named. The *Bithorax* gene complex, for example, controls the identity of the third segment of the fly's thorax. The *Antennapedia* gene complex controls the identity of the head and thorax segments.

Exactly how all these genes work is not yet known. But we do know that the homeotic genes encode proteins that can turn other genes ON or OFF. This ability allows them to exercise powerful control over parts of the development process. The homeotic genes can regulate themselves and each other. Through positive and negative feedback, they reinforce themselves and repress the others. As a result, in any one cell only one homeotic gene can be stably expressed. This feature ensures the stable expression of the segments.

The homeotic genes in many species have all been found to have in them a sequence of 180 nucleotides that are very much alike from one gene to another. This sequence has been called the *homeobox*. The name is taken after the homeotic genes. Despite the name, the homeobox has been found also in genes other than the homeotic genes.

---

*I shall not explain the meaning of the names given to these genes. Many of their names derive from aberrant effects of mutations in them. The names of these genes can be confusing to one who is trying to understand their normal function.

A gene containing a homeobox encodes a regulatory protein [Marx 1988, Robertson 1988, De Robertis et al. 1990]. The 180 nucleotides of the homeobox encode a 60-amino-acid segment of that protein. This segment of the regulatory protein is called the *homeodomain*. The homeodomain in these proteins has been found to bind to the DNA of a target gene. From that position the protein activates or represses the gene. Molecular biologists have found homeoboxes in all species they've studied. They've found them in worms, flies, frogs, mice, and humans. They've also found them in plants [Vollbrecht et al. 1991]. They suspect all animals and plants have homeoboxes.

Embryologists have long known that during development some tissues act as *inductors* to neighboring tissues. That means that these tissues can, by their presence, cause their neighbor cells to pass from one state to another. If the inductor is absent, the cell will not change into the proper state. Scientists have now found some of the molecular messages these inductors send out to make the induction happen.

The induction of a cell in the photoreceptor of the *Drosophila* eye is a well-studied example. There are eight cells in the photoreceptor, denoted by R1 through R8. The cell R8 is adjacent to R7 and induces its development. The way it does that has been found to be as follows. The cell R8 has a gene called the *boss* gene that encodes a message protein called the *boss* protein. The boss protein sits on the surface of cell R8 at the point where it touches cell R7. At the contact point on the surface of the proto-R7 cell there is a receptor protein called *sev* (for the *seven* of R7).

The *sev* protein sits within the membrane of the cell that will become R7. Like Janus, the *sev* protein looks with one face toward the outside of the cell and with another face it looks toward the inside. *Sev* is therefore called a *transmembrane receptor*. The *boss* protein binds to *sev* and is captured by the proto-R7 cell. On the inside end of the *sev* protein, is an active site for an enzyme called a *protein tyrosine kinase*.

A protein kinase is an enzyme that adds a phosphate group of atoms (phosphorous with oxygen) to a special site on a target enzyme. This process is called *phosphorylation*, and leads to an activation of the target enzyme. Actually, a single protein kinase can activate many molecules of target enzyme, resulting in an amplification of the signal. This cascade of activity induces further enzyme activity that signal gene changes that turn the proto-R7 cell into a full-fledged R7. All of this activity is an example of what is called *inductive interaction* [Griffiths et al. 1993].

J.   A HERITABLE GENETIC SWITCH

Here I shall describe in some detail the genetic switch I suggested in Chapter 7. The switch is arranged as a flip-flop, and anyone familiar with a flip-flop in electronic circuitry will recognize it immediately. The switch comprises two genes which try to repress each other. Mutual repression of genes is known to occur [Weigel and Meyerowitz 1993]. The switch is stable in either of two states. It will remain in whatever state it finds itself until it is triggered to flip to the other state.

Fig. A.13 shows a pair of genes, labeled A and B, arranged as a binary switch. The line coming out the top of each gene in the figure is its output, and represents those proteins whose synthesis is caused by the gene. We might think of each of these genes as controlling an array of other genes. These other genes might encode some function like a set of enzymes catalyzing a set of biochemical reactions. In a multicelled organism these genes might turn ON a subprogram in the development of the embryo.

The two lines coming into the gene from below are its control signals. The signal line terminating on the gene with an arrowhead is an inducer, and that terminating with a small open circle is a repressor. An environmental cue I shall call E_A activates gene A through the inducer input, and an environmental cue I shall call E_B activates gene B. The cues can

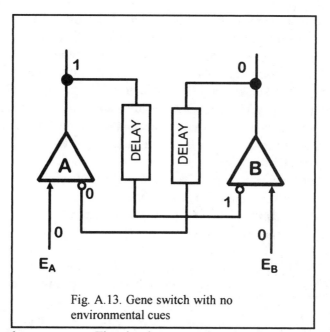

Fig. A.13. Gene switch with no
environmental cues

be only ON or OFF. That is, they can take on only two values:
ZERO and ONE. I define the cues in such a way that they cannot both be ON at the same time. That is to say, I define the
environments to be mutually exclusive.

In the figure, the output of a gene is labeled 0 (ZERO) if the
gene is OFF, and 1 (ONE) if the gene is ON. If the repressor input to a gene is 1, the gene is turned OFF. If the inducer is 1,
it overrides the repressor and turns the gene ON. (The inducer
actually removes the repressor from the operator section of
the gene.) A gene in the figure is ON if its repressor is 0 *or* if
its inducer is 1. It is OFF if its repressor is 1 *and* its inducer is
0. The switch is bistable. It is stable in either of two states.
One state is with *A* ON and *B* OFF. The other state is the opposite, with *B* ON and *A* OFF. Only one gene can be ON at a
time. In Fig. A.13 gene *A* is ON and *B* is OFF.

When one of the genes is ON, it not only activates the
function it controls, it also causes a protein to be made that
represses the other gene. Fig. A.13 shows the repressor to

each gene coming from the output of the other gene. Each gene tries to repress the other, as is indicated by the line from the output of one gene to the repressor of the other. Each of these repressor signals is shown going through a delay. The role of these delays will become evident shortly. If we at first think of the genetic switch in steady state, then the role of the delays becomes irrelevant and can be ignored. The delays play a role only in the transient stage right after the genes have switched states.

Proteins in a cell that have ceased to be useful do not degrade spontaneously. Peptide bonds, if left alone, can remain intact for months or even years. Cells break down their inactive proteins only when they have a reason to do so. And when they do, they do it in a controlled way [Stryer 1988, p. 794]. Unless the cell actively breaks them down, protein molecules remain intact. They remain in the cell until they are diluted out through cell division. The genetic-switch model calls for the cell to start degrading a repressor protein as soon as the gene that commands its synthesis is shut OFF. The delay in the disappearance of the repressor is represented by the delay boxes in the figures.

When a gene that's ON gets shut OFF, it turns OFF the function it controls. The gene also stops commanding the making of the repressor protein. But the repressor molecules that have already been made will still be in the cell. The presence of both repressors will keep both genes OFF, and this state of the switch is unstable.

To maintain stability in the switch, the model requires these repressor proteins to be actively degraded when they are both present. The presence of both repressors might, for example, provide a signal to the cell to start degrading both of them. Cells are known to have the means to degrade protein selectively [Gottesman 1989].

The inducer inputs are cues from the environment. Note that if E_A is ON, it induces gene A to turn ON. If E_B is ON, it induces gene B to turn ON. The cues E_A and E_B cannot both be ON at the same time. At least one of them must be OFF. They might both be OFF.

Fig. A.13 shows a condition in which both environmental cues E_A and E_B are OFF. The state of the switch shown in Fig. A.13 is stable because the output of A represses B and keeps it OFF. Gene A is ON because its repressor input is 0. This state of the switch, in which A is ON and B is OFF, will maintain itself even when there is no cue from the environment.

The state of the switch is naturally passed on to the next generation, and is therefore heritable. The state will maintain itself indefinitely through innumerable cell divisions. When the cell divides, the repressor protein gets divided between the daughter cells. But it will not be diluted, because the daughter cells will continue to make it. They will continue to make it because there is no repressor to gene A, and the gene will be ON in the daughter cells after the cell divides. As long as gene A is ON, it keeps gene B OFF. As long as B is OFF, A will be ON. Gene A will continue to make the repressor for gene B and build up the supply in each daughter cell to the level it was in the mother cell. This situation will persist as long as there are no environmental cues. This process will re-

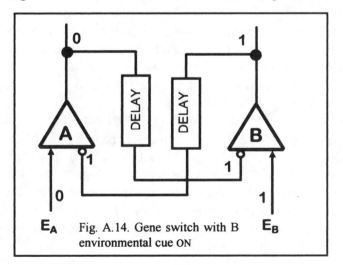

Fig. A.14. Gene switch with B environmental cue ON

peat with each cell division. And it is this process that makes the state heritable.

In Fig. A.13 there are no cues from the environment. The environment is neither E_A nor E_B. It is a neutral environment in which the switch can be stable in either state. Now let the environment change from neutral to E_B so that the E_B input to gene *B* becomes 1, as shown in Fig. A.14. Although gene *B* still has a repressor input of 1, the inducer overrides it and turns the gene ON. When gene *B* is switched ON, it makes a repressor for gene *A*. Gene *A* is then turned OFF and stops making the repressor to *B*. Note that although gene *A* has stopped making the repressor for *B*, the repressor it had already made remains in the cell until it gets diluted out through cell division. The switch has now flipped to its other stable state. In this state, gene B is ON and gene A is OFF. This state also will be passed on to the next generations and will be heritable.

Suppose now the environment changes back to neutral. E_B goes back to 0. What happens next depends on how long gene *B* has been ON. If *B* has been ON (and *A* has been OFF) through sufficiently many cell generations for the repressor made by *A* to have diluted out, then the repressor to *B* will be 0.

In this case *B* remains ON as shown in Fig. A.15. The system retains the state induced when E_B became 1. The induced state continues in the cell and is passed on to later generations. The new traits the cell acquired when E_B became 1 are also passed on to later generations.

But suppose E_B was ON for only a short time. Suppose it changed back to 0 *before* the repressor made by *A* had diluted out. Then even though the output of *A* is 0, the repressor input to *B* is still 1, as shown in Fig. A.16. Now when E_B becomes 0, *B* shuts OFF. Both genes are OFF, and the switch is in an unstable state. With both *A* and *B* OFF, the repressor-degrading function is activated and the repressors to both *A* and *B* begin to be degraded.

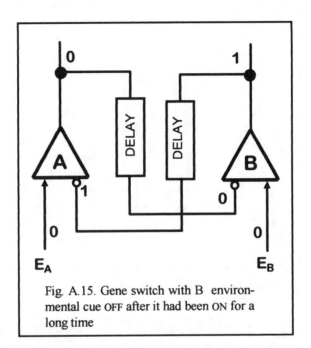

Fig. A.15. Gene switch with B environmental cue OFF after it had been ON for a
long time

What happens next depends on which of the two repressors
disappears first. Which repressor disappears first determines
which gene goes ON.

If the environmental cue E_B was ON for only a short time,
the change it made will not pass on to the next generation and
the switch will revert to its former state. If it was ON for a
long enough time, the change will maintain itself into future
generations.

This model explains the dauer effect first reported by Vic-
tor Jollos [1921], then at the Kaiser-Wilhelm Institute for
Biology in Germany. Jollos coined the German term *Dauer-
modifikationen*, meaning *long-lasting modifications*, to describe heritable changes induced in paramecia by the environment. The terms *dauer modification*, and *dauer effect*,
have since become accepted English terms for this phenomenon.

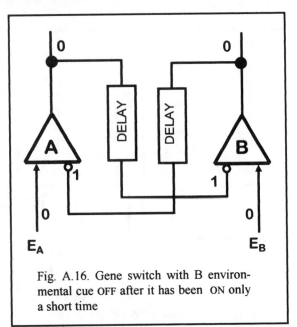

Fig. A.16. Gene switch with B environmental cue OFF after it has been ON only a short time

In the model of the heritable switch, a change in the environment sends a signal to each member of the population. This signal activates a genetic command in each individual to call up a preprogrammed subroutine. If the new environment does not last long, the population will revert to its previous state. But if the new environment persists for a long enough time, then even after the environment changes, the population will remain in its new state. The new state is carried into future generations, and to this extent it is heritable. The heritability is, however, not absolute. A different cue can make the population change again. But the longer the environmental cue lasts, the more nearly heritable the effect appears.

REFERENCES[*]

Agard, D. A., (1993). "To fold or not to fold," *Science*, vol. 260, pp. 1903-1904.

Aharonowitz, Y. and G. Cohen, (1981). "The microbiological production of pharmaceuticals," *Scientific American*, vol. 245, May.

Amábile-Cuevas, C. F., M. Cárdenas-Garcia, and M. Ludgar, (1995). "Antibiotic resistance," *American Scientist*, vol. 83, pp. 320-329.

Applebury, M. L., (1990). "Insight into blindness," *Nature*, vol. 343, pp. 316-317.

Ayala, F. J., (1978). "The mechanisms of evolution," *Scientific American*, vol. 239, September.

Beeman, R. W., (1982). "Recent advances in the mode of action of insecticides," *Annual Review of Entomology*, vol. 27, pp. 253-281.

Bahill, A. T. and L. Stark, (1979). "The trajectories of saccadic eye movements," *Scientific American*, vol. 240, January.

Benson, S. A., (1988). "Is bacterial evolution random or selective?", *Nature*, vol. 336, pp. 21-22.

Bergerud A. T., (1983). "Prey switching in a simple ecosystem," *Scientific American*, vol. 249, December.

---

[*] For *Scientific American* no pagination is given to avoid confusion. The pagination of the US and the foreign editions of *SA* differ. Instead the month of the issue is given and the reference can be found easily.

Bishop, J. A. and L. M. Cook, (1975). "Moths, melanism and clean air," *Scientific American*, vol. 232, January.

Bone, R., J. L. Silen, and D. A. Agard, (1989). "Structural plasticity broadens the specificity of an engineered protease," *Nature*, vol. 339, pp. 191-195

Bonner, J. T., (1988). *The Evolution of Complexity by Means of Natural Selection*, Princeton: Princeton University Press.

Boyajian, G. and T. Lutz, (1992). "Evolution of biological complexity and its relation to taxonomic longevity in the Ammonoidea," *Geology*, vol. 20, pp. 983-986.

Bradshaw, A. D., (1965). "Evolutionary significance of phenotypic plasticity in plants," *Advances in Genetics*, vol. 13, pp. 115-155.

Brock, R. D. (1980). "Mutagenesis and Crop Production," in Carlson, P. S., *The Biology of Crop Productivity*, New York: Academic Press pp. 383-409.

Brooks J. L., (1957). In Mayr [1957], pp. 81-123.

Burleigh, B. D., Rigby, P. W. J., Hartley, B. S., (1974). "A comparison of wild-type and mutant ribitol dehydrogenase from *Klebsiella aerogenes*." *Biochemical Journal* vol. 143, pp. 341-352.

Cairns-Smith, A. G. (1984). *Genetic Takeover and the Mineral Origins of Life*, Cambridge: Cambridge University Press.

Cairns-Smith, A. G. (1985). "The first organisms," *Scientific American*, vol. 252, June.

Cairns, J., J. Overbaugh, and S. Miller, (1988). "The origin of mutants," *Nature*, vol. 335, pp. 142-145.

Campbell, J. H., J. A. Lengyel, and J. Langridge, (1973). "Evolution of a second gene for β-galactosidase in *Escherichia coli*," *Proceedings National Academy of Science USA*, vol. 70, pp. 1841-1845.

Cannon, H. Graham, (1958). *The Evolution of Living Things*, Manchester: University of Manchester Press.

Carroll, R. L., (1988). *Vertebrate Paleontology and Evolution*, New York: W. H. Freeman.

Collier, B. D., G. W. Cox, A. W. Johnson, and P. C. Miller, (1973). *Dynamic Ecology*, Englewood Cliffs: Prentice-Hall.

Conant, S., (1988). "Saving endangered species by translocation," *BioScience*, vol. 38, pp. 254-257

Cook, L. M., G. S. Mani, and M. E. Varley, (1986) "Postindustrial Melanism in the Peppered Moth," *Science*, vol. 231, pp. 611-613.

Cook, N. D., (1977). "The case for reverse translation," *Journal of Theoretical Biology*, vol. 64, pp. 113-135.

Craig, E. A., (1993). "Chaperones: helpers along the pathways to protein folding," *Science*, vol. 260, pp. 1902-1903.

Culotta, E., (1994). "A boost for 'adaptive' mutation," *Science*, vol. 265, pp. 318-319.

Darnell, J. E., H. F. Lodish, and D. Baltimore, (1986). *Molecular Cell Biology*. New York: Scientific American Books.

Darwin, C. G., (1959). "Some Episodes in the Life of Charles Darwin," Proceedings *American Philosophical Society*, vol. 103, pp. 609-615.

Darwin, C., (1872). *The Origin of Species*. Reprint of sixth edition (1963). New York: Dutton.

Davies, J. and M. Nomura, (1972). "The genetics of bacterial ribosomes," *Annual Review of Genetics*, vol. 6, pp. 203-234.

Davies, J., M. Brzezinska, and R. Benveniste, (1971). "R factors: Biochemical mechanisms of resistance to aminoglycoside antibiotics," *Annals of the NY Academy of Science*, vol. 182 pp. 226-233.

Dawkins, R. (1986). *The Blind Watchmaker*. New York & London: W. W. Norton.

De Robertis, E. M., G. Oliver, and C. V. E. Wright, (1990). "Homeobox genes and the vertebrate body plan," *Scientific American*, vol. 263, July.

Delage, Yves, (1903). *L'hérédité*. (As cited by Thompson, D'Arcy [1942].)

Dobzhansky, T., (1951). *Genetics and the Origin of Species*, 3rd ed. New York: Columbia University Press.

Donelson, J. E., and M. J. Turner, (1985). "How the trypanosome changes its coat," *Scientific American*, vol. 252, February.

Drake, J. W., (1969). "Comparative rates of spontaneous mutation," *Nature*, vol. 221, p. 1132.

Drake, J. W., (1991). "Spontaneous mutation," *Annual Reviews of Genetics*, vol. 25, pp. 125-146.

Drummond, H., (1894). *The Ascent of Man*, London (as cited by Himmelfarb [1962, p. 393]).

Dryja, T. P., T. L. McGee, E. Reichel, L. B. Hahn, G. S. Cowley, D. W. Yandell, M. A. Sandberg, and E. L. Berson, (1990). "A point mutation of the rhodopsin gene in one form of retinitis pigmentosa," *Nature*, vol. 343, pp. 364-366.

Eldredge, N. and S. J. Gould, (1972). "Punctuated equilibria: an alternative to phyletic gradualism", in Schopf, T. J. M. (ed.). *Models in Paleobiology*, Freeman, Cooper and Co., San Francisco, pp. 82-115.

Eldredge, N. and S. J. Gould, (1988). *Nature*, vol. 332, pp. 211-212.

Eliot, C. W., (1909). "Introductory Note" to Darwin's *Origin of Species*, Harvard Classics edition, Volume 11, New York: Collier.

Fedoroff, N. V. (1984). "Transposable genetic elements in maize", *Scientific American*, vol. 250, June.

Fedoroff, Nina and David Botstein (eds.), (1992). *The Dynamic Genome: Barbara McClintock's Ideas in the Century of Genetics*, Cold Spring Harbor Laboratory Press.

Feller, W., (1957). *An Introduction to Probability Theory and Its Applications*, Second edition (first edition published 1950), New York: Wiley.

Fersht, A. R. (1981). "DNA replication fidelity," *Proceedings of the Royal Society (London)*, vol. B 212, pp. 351-379.

Fischbach, G. D., (1992). "Mind and Brain," *Scientific American*, vol. 267, September.

Fisher, R. A. (1958). *The Genetical Theory of Natural Selection*, Oxford. Second revised edition, New York: Dover.

Foster, P. L., (1992). "Directed mutation: between unicorns and goats," *Journal of Bacteriology*, vol. 174, pp. 1711-1716.

Foster, P. L. and J. M. Trimarchi, (1994). "Adaptive reversion of a frameshift mutation in *Escherichia coli* in simple base deletions in homopolymeric runs," *Science*, vol. 265, pp. 407-409.

Frost H. M., (1973). *Bone Remodeling and Skeletal Modeling Errors*, Springfield: C. H. Thomas.

Fryer, G., P. H. Greenwood, and J. F. Peake, (1983). "Punctuated equilibria, morphological stasis and the palaeontological documentation of speciation: a biological appraisal of a case history in an African lake," *Biological Journal of the Linnean Society*, vol. 20, pp. 195-205.

Galef, B. G., (1985). In Johnston and Pietrewicz [1985], pp. 143-166.

Galitski,T. and R. R. Roth, (1995). "Evidence that F plasmid transfer replication underlies apparent adaptive mutation," *Science*, vol. 268, pp. 421-423.

Gartner, T. K. and E. Orias, (1966). "Effects of mutations to streptomycin resistance on the rate of translation of mutant genetic information," *Journal of Bacteriology*, vol. 91, pp. 1021-1028

Gilbert, W., (1978). "Why genes in pieces?", *Nature*, vol. 271, p. 501.

Glass, Bentley, (1965a). "The ethical basis of science," *Science*, vol. 150, pp. 1254-1261.

Goldschmidt, R. B., (1940). *The Material Basis of Evolution*, New Haven. Yale University Press.

Goldschmidt, R. B., (1952). "Evolution as viewed by one geneticist," *American Scientist*, vol. 40, 135, pp. 84-98.

Gottesman S. (1989). "Genetics of proteolysis in Escherichia Coli," *Annual Review of Genetics*, vol. 23, pp. 163-198.

Gould, S. J., (1977). "Evolution's erratic pace," *Natural History*, vol. 8, pp. 12-16.

Gould, S. J., (1980). "Is a new and general theory of evolution emerging?" *Paleobiology*, vol. 6, pp. 119-130.

Gould, S. J. (1989). *Wonderful Life: The Burgess Shale and the Nature of History*, 347 pp. WW Norton.

Gould, S. J. and N. Eldredge, (1977). "Punctuated equilibria: the tempo and mode of evolution reconsidered," *Paleobiology*, vol. 3, pp. 115-151.

Grant, P. R., (1986). *Ecology and Evolution of Darwin's Finches*, Princeton: Princeton University Press.

Griffiths, A. J. F., J. H. Miller, D. T. Suzuki, R. C. Lewontin, and W. M. Gelbart, (1993). *An Introduction to Genetic Analysis*, New York: Freeman.

Grosse, F., G. Krauss, J. W. Knill-Jones, and A. R. Fersht, (1984). "Replication of ϕX174 DNA by calf thymus DNA polymerase α: Measurement of error rates at the amber-16 codon," in *Advances in Experimental Medicine and Biology, vol. 179, Proteins Involved in DNA Replication*, pp. 535-540.

Hall, B. G., (1978). "Regulation of newly evolved enzymes. IV. Directed evolution of the eβg repressor," *Genetics*, vol. 90, pp. 673-681.

Hall, B. G., (1982). "Evolution in a Petri dish: The evolved β-galactosidase system as a model for studying acquisitive evolution in the laboratory," *Evolutionary Biology*, vol. 15, pp. 85-150.

Hall, B. G., (1988). "Adaptive evolution that requires multiple spontaneous mutations. I. Mutations involving an insertion sequence," *Genetics*, vol. 120, pp. 887-897.

Harrison R. G., (1980). "Dispersal polymorphisms in insects, " *Annual Reviews of Ecological Systems*, vol. 11, pp. 95-118.

Hartl, D. L., and B. G. Hall, (1974). "Second naturally occurring β-galactosidase in E. Coli," *Nature*, vol. 248, pp. 152-153.

Hermas, S. A., C. W. Young, and J. W. Rust (1987). "Genetic relationships and additive genetic variation of productive and reproductive traits in Guernsey dairy cattle," *Journal of Dairy Science*, vol. 70, pp. 1252-1257.

Herring S. W. and T. C. Lakars (1981). *Journal of Craniofacial Genetics and Development Biology*, vol. 1, p. 341.

Herskowitz, I. H., (1962). *Genetics*, Boston: Little, Brown.

Herskowitz, I. H. and D. Hagen, (1980). "The lysis-lysogeny decision of phage lambda: explicit programming on responsiveness." *Annual Review of Genetics*, vol. 14, pp. 399-446.

Himmelfarb, G., (1962). *Darwin and the Darwinian Revolution*, Garden City: Doubleday

Hinegardner, R. and J. Engleberg, (1983). "Biological complexity," *Journal of Theoretical Biology*, vol. 104, pp. 7-20.

Ho, M -W. and P. T. Saunders, (1979). "Beyond Neo-Darwinism An epigenetic approach to evolution," *Journal of Theoretical Biology*, vol. 78. p. 573-591.

Holliday, Robin, (1989). "A different kind of inheritance," *Scientific American*, vol. 260 June.

REFERENCES

Hoyle, F. and N. C. Wickramasinghe, (1982). *Why Neo-Darwinism Does Not Work*, Cardiff: University College Cardiff Press.

Hubbs C. L., (1922). "Variation in the number of vertebrae and other meristic characters of fishes correlated with the temperature of water during development," *American Naturalist*, vol. 56, p. 360-372.

Hubbs C. L., (1926). "The structural consequences of modification of the developmental rate in fishes, considered in reference to certain problems in evolution," *American Naturalist*, vol. 60, p. 57-81.

Hubbs, C. L., (1967). "Electric eel," *Encyclopædia Britannica*, vol. 8, p. 127.

Hubel, D. H., (1963). "The visual cortex of the brain," *Scientific American*, vol. 209, November.

Huxley, J., (1943). *Evolution: The Modern Synthesis*, New York and London: Harper.

Inderlied, C. B. and R. P. Mortlock, (1977). "Growth of *Klebsiella aerogenes* on xylitol: implications for bacterial enzyme evolution," *Journal of Molecular Evolution*, vol. 9, pp. 181-190.

Johnston, T. D. and G. Gottlieb, (1990). "Neophenogenesis: A developmental theory of phenotypic evolution," *Journal of Theoretical Biology*, vol. 147, pp. 471-495

Johnston, T. D. and A. T. Pietrewicz (eds.), (1985). *Issues in the Ecological Study of Learning*, Hillsdale, NJ: Erlbaum Assoc.

Jollos, V., (1921). "Experimentelle Protistenstudien, I. Untersuchungen über Variabilität und Vererbung bei Infusorien," *Archiv für Protistenkunde*, vol. 43, pp. 1-222.

Kalat J. W., (1985). In Johnston and Pietrewicz [1985], pp. 119-141.

Katz, M. J., (1987). "Is evolution random?" In Raff and Raff, pp. 285-315.

Kettlewell, H. B. D., (1955). "Selection experiments on industrial melanism in the Lepidoptera," *Proceedings Royal Society*, B, vol. 145, pp. 297-303.

Kettlewell, H. B. D., (1956). "A resume of investigations on the evolution of melanism in the Lepidoptera," *Proceedings Royal Society*, B, vol. 145, pp. 297-303.

Kettlewell, H. B. D., (1957). "The contribution of industrial melanism in the Lepidoptera to our knowledge of evolution," *The Advancement of Science*, vol. 13, pp. 245-252.

Kettlewell, H. B. D., (1959). "Darwin's missing evidence," *Scientific American*, vol. 200, March.

Kettlewell, H. B. D., (1973). *The Evolution of Melanism: The Study of a Recurring Necessity with Special Industrial Melanism in the Lepidoptera*, Oxford University Press.

Kohler, I., (1962). "Experiments with goggles," *Scientific American*, vol. 206, May.

Konzak, C. F. (1977). "Genetic control of the content, amino acid composition and processing properties of proteins in wheat," *Advances in Genetics*, vol. 19, pp. 407-582.

Kraut, J. (1988). "How do enzymes work?" *Science*, vol. 242, pp. 533-540.

Landau, R. L., (1967). "Endocrinology," *Encyclopædia Britannica*, vol. 8, pp. 378-382.

Lanyon L. E., (1980). "The influence of function on the development of bone curvature. An experimental study on the rat tibia," *Journal of Zoology*, vol. 192, pp. 457-466.

Lenski, R. E. and J. E. Mittler, (1993). "The directed mutation controversy and neo-Darwinism," *Science*, vol. 259, pp. 188-194.

Lerner, S. A., T. T. Wu, and E. C. C. Lin, (1964). "Evolution of a catabolic pathway in bacteria," *Science*, vol. 146, pp. 1313-1315.

REFERENCES

Levine, J. S., and E. F. MacNichol, Jr., (1982). "Color vision in fishes," *Scientific American*, vol. 246, February.

Lewontin, R. C., (1978). "Adaptation," *Scientific American*, vol. 239, September.

Linder, M. E. and A. G. Gilman, (1992). "G proteins," *Scientific American*, vol. 267, July.

Lively, C. M., (1986). "Predator-induced shell dimorphism in the acorn barnacle Chthamalus-anisopoma," *Evolution*, vol. 40, pp. 232-242.

Luria, D., (1990) כתבי הגאון ר׳ דוד לוריא, הפירוש על פרקי דרבי אליעזר, Jerusalem: Zonnenfeld. (New revised edition).

Luria, S. E. & M. Delbrück, (1943). "Mutation of bacteria from virus sensitivity to virus resistance," *Genetics*, vol. 28, 491-511.

Luzzatto, L. and P. Goodfellow, (1989). "Sickle-cell anemia A simple disease with no cure," *Nature*, vol. 337, pp. 17-18.

Marx, J. L., (1988). "Homeobox linked to gene control," *Science*, vol. 242, pp. 1008-1009.

Maynard Smith, J., (1988). "Punctuation in perspective," *Nature*, vol. 332, pp. 311-312.

Mayr, E. (ed.), (1957). *The Species Problem*, Washington DC: AAAS.

McClintock, B., (1978). "Mechanisms that rapidly reorganize the genome," reprinted in Fedoroff and Botstein [1992].

McCoy, J. W., (1977). "Complexity in organic evolution," *Journal of Theoretical Biology*, vol. 68, pp. 457-458.

McDonald, J. F., (1990). "Macroevolution and retroviral elements," *BioScience*, vol. 40, pp. 183-191.

McDonald. J. F., G. K. Chambers, J. David, and F. J. Ayala,(1977). Adaptive response due to changes in gene regulation: a study with *Drosophila*," *Proceedings National Academy of Sciences, USA*, vol. 74, pp. 4562-4566.

McShea, D. W., (1991). "Complexity and evolution: What everybody knows," *Biology and Philosophy*, vol. 6, pp. 303-324.

McShea, D. W., (1993). "Evolutionary change in the morphological complexity in the mammalian vertebral column," *Evolution*, vol. 47, pp. 730-740.

Michael, C. R., (1969). "Retinal processing of visual images," *Scientific American*, vol. 220, May.

Moorehead, P. S. and M. M. Kaplan (eds.), (1967). *The Mathematical Challenges to the Neo-Darwinian Interpretation of Evolution*, Philadelphia: Wistar Institute. of Anatomy and Biology.

Mortlock, R. P., (1982). "Metabolic acquisition through laboratory selection," *Annual Review of Microbiology*, vol. 36, pp. 259-284.

Murray, A. W. and M. W. Kirschner (1991). "What controls the cell cycle," *Scientific American*, vol. 264, March.

Murray C. B. and T. D. Beacham (1989). "Responses of meristic characters in chum salmon (Oncorhynchus keta) to temperature changes during development," *Canadian Journal of Zoology*, vol. 67, pp. 596-600.

Needham, A. E., (1965). *The Uniqueness of Biological Materials*, London: Pergamon.

Nichols, J. T. and C. Hubbs, (1967). "Fish," *Encyclopædia Britannica*, vol. 9, p. 332.

Normark, S., S. Bergström, T. Edlund, T. Grundström, B. Jaurin, F. P. Lindberg, and O. Olsson, (1983). "Overlapping genes," *Annual Review of Genetics*, vol. 17, pp. 499-525.

Novacek, M. J., (1985). "Evidence for echolocation in the oldest known bats," *Nature*, vol. 315, pp. 140-141.

O'Farrell, P. H., B. A. Edgar, D. Lakich, and C. F. Lehner, (1989). "Directing cell division during development," *Science*, vol. 246, pp. 635-640.

Packard, A., (1972). "Cephalopods and fish: The limits of convergence," *Biological Reviews*, vol. 47, pp. 241-307.

Paley, W., (1817). *Natural Theology: or Evidences of the Existence and Attributes of the Deity Collected from the Appearances of Nature*, London: Rayer.

Persson, G. and A. Hagberg, (1969). "Induced variation in a quantitative character in barley: morphology and cytogenetics of erectoides mutants," *Hereditas*, vol. 61, pp. 115-178.

Pettigrew, J. D., 1972 "The neurophysiology of binocular vision, " *Scientific American*, vol. 227, August.

Pimm, S. L., (1988). "Rapid morphological change in an introduced bird," *Trends in Evolution and Ecology*, vol. 3, pp. 290-291.

Quiring, R., U. Walldorf, U. Kloter, and W. J. Gehring, (1994). "Homology of the eyeless gene of *Drosophila* to the Small eye gene in mice and Aniridia in humans," *Science*, vol. 265, pp. 785-789.

Radicella, J. P., P. U. Park, and M. S. Fox, (1995). "Adaptive mutation in *Escherichia coli*: A role for conjugation," *Science*, vol. 268, pp. 418-420.

Raven, C. E. (1967). "William Paley," *Encyclopædia Britannica*.

Rensch, B., (1960). *Evolution Above The Species Level*, New York: Columbia University Press (2nd ed.).

Reznick, D. A. and H. Bryga, (1987). "Life-history evolution in guppies (Poecilia reticulata): 1. Phenotypic and genetic changes in an introduction experiment," *Evolution*, vol. 41, pp. 1370-1385.

Reznick, D. A., H. Bryga, and J. A. Endler, (1990). "Experimentally induced life-history evolution in a natural population," *Nature*, vol. 346, pp. 357-359.

Reznick, D. A., F. H. Shaw, F. H. Rodd, & R. G. Shaw, (1997) "Evaluation of the rate of evolution in natural populations of guppies (*Poecilia reticulata*). *Science*, vol. 275, pp. 1934-1937.

Richter C. P., (1947). *Journal of Comparative Physiology and Psychology*, vol. 40, p. 129.

Rigby, P. W. J., B. D. Burleigh, and B. S. Hartley, (1974). "Gene duplication in experimental enzyme evolution," *Nature* vol. 251, pp. 200-204.

Robertson, M. (1988). "Homeoboxes, POU proteins and the limits to promiscuity," *Nature*, vol. 336, pp. 522-524.

Rosen, Donn E. and Donald G. Buth (1980). "Empirical evolutionary research versus neo-Darwinian speculation," *Systematic Zoology*, vol. 29, pp. 300-308.

Rosenberg, S. M., S. Longerich, P. Gee, and R. S. Harris, (1994). "Adaptive mutation by deletions in small mononucleotide repeats, *Science*, vol. 265, pp. 405-407.

Rowland, M. W. (1987). "Fitness of insecticide resistance," *Nature*, vol. 327, p. 194.

Ruse. M. (1982). *Darwinism Defended*, Reading: Addison-Wesley.

Rushton, W. A. H., (1962). "Visual pigments in man," *Scientific American*, vol. 207, November.

Savageau, M. A., (1976). *Biochemical Systems Analysis. A Study of Function and Design in Molecular Biology*, Reading: Addison-Wesley.

Schindewolf, O. H., (1950). *Grundfragen der Paläontologie*, Stuttgart. Schweizerbart.

Schlichting, C. D., (1986). "The evolution of phenotypic plasticity in plants," *Annual Review of Ecological Systems*, vol. 17, pp. 667-693.

Schnapf, J. L. and D. A. Baylor, (1987). "How photoreceptor cells respond to light," *Scientific American*, vol. 256, April.

REFERENCES

Schreider, E., (1964). "Ecological rules, body-heat regulation, and human evolution," *Ecology*, vol. 18, pp. 1-9.

Shapiro, J. A., (1995). "Adaptive mutation: Who's really in the garden?" *Science* vol. 268, pp. 373-374.

Silverman, M., and M. Simon, (1983). "Phase variation and related systems," in Shapiro J. A. (ed.), (1983). *Mobile Genetic Elements*, New York: Academic Press, 537-557.

Simpson, G. G., (1953). *The Major Features of Evolution*, New York: Columbia University Press.

Simpson, G. G., (1960). "The world into which Darwin led us," *Science*, vol. 131, pp. 966-974.

Simpson, G. G., (1961). *Horses*, Garden City: Doubleday.

Soller, M. and J. S. Beckmann, (1987). "Cloning quantitative trait loci by insertional mutagenesis," *Theoretical and Applied Genetics*, vol. 74, pp. 369-378.

Smith, T. B., (1987). "Bill size polymorphism and intraspecific niche utilization in an African finch," *Nature*, vol. 329, pp. 717-719.

Sniegowski, P. D. & R. E. Lenski (1995). "Mutation and adaptation: The directed mutation controversy in evolutionary perspective." *Annual Review of Ecology and Systematics* vol. 26 pp. 553-578.

Spetner, L. M. (1964). "Natural selection: an information-transmission mechanism for evolution," *Journal of Theoretical Biology*, vol. 7, pp. 412-419.

Spetner, L. M. (1966). "Mutation — the pacemaker of evolution," *Proceedings 2nd International Congress on Biophysics*, Vienna.

Spetner, L. M. (1968). "Information transmission in evolution," *IEEE Transactions on Information Theory*, vol. IT-14, pp. 3-6.

Spetner. L. M. (1970). "Natural selection versus gene uniqueness," *Nature*, vol. 226. pp. 948-949.

Stahl, F. W., (1987). "Genetic recombination," *Scientific American*, vol. 256, February.

Stahl, F. W., (1988). "A unicorn in the garden," *Nature*, vol. 335, pp. 112-113.

Stanley, S. M., (1979). *Macroevolution: Pattern and Process*, San Francisco: Freeman.

Stearns, S. C., (1989). "The evolutionary significance of phenotypic plasticity," *BioScience*, vol. 39, pp. 436-445.

Stebbins, G. L. (1966). *Processes of Organic Evolution*, Englewood Cliffs: Prentice-Hall.

Stewart, C. -B., J. W. Schilling, and A. C. Wilson, (1987). "Adaptive evolution in the stomach lysozyme of foregut fermenters," *Nature*, vol. 330, pp. 401-404.

Stryer, L., (1988). *Biochemistry*, (3rd ed.) New York: Freeman.

Sumner F. B., (1909). "Some effects of external conditions on the white mouse," *Journal of Experimental Zoology*, vol. 7, pp. 97-155.

Tanaka, K., J. G. Scott, F. Matsumura, (1984). "Picrotoxinin receptor in the central nervous system of the American cockroach: its role in the action of cyclodiene-type insecticides," *Pesticide Biochemistry and Physiology*, vol. 22, pp. 117-127.

Thompson, D'Arcy W., (1942). *On Growth and Form*: A New Edition, Cambridge: Cambridge University Press. Republished 1992, New York: Dover.

Thompson, L. W. and S. Krawiec, (1983). "Acquisitive evolution of ribitol dehydrogenase in *Klebsiella pneumoniae*," *Journal of Bacteriology*, vol. 154, pp. 1027-1031

Turesson, G., (1925). "The plant species in relation to habitat and climate," *Hereditas*, vol. 6, pp. 147-236,

Vollbrecht, E., B. Veit, N. Sinha, and S. Hake, (1991). "The developmental gene Knotted-1 is a member of a maize homeobox gene family," *Nature*, vol. 350, pp. 241-243.

Walbot, V. and Cullis, C. A. (1985). "Rapid genomic change in higher plants," *Annual Review of Plant Physiology*, vol. 36, pp. 367-396.

Wanner, B. L., (1985). "Phase Mutants: Evidence of a physiologically regulated 'change-in-state' gene system in *Escherichia coli*," in Simon, M. and I. Herskowitz (eds.), *Genome Rearrangement, Proceedings of the UCLA Symposium Apr. 7-13 1984*, New York: Alan R. Liss, pp. 103-122.

Watson, J. D. and F. H. C. Crick, (1953a). "Molecular structure of nucleic acids. A structure for deoxyribose nucleic acid," *Nature*, vol. 171. p. 737-738.

Watson, J. D. and F. H. C. Crick, (1953b). "The structure of DNA," *Cold Spring Harbor Symposium on Quantitative Biology*, vol. 18, pp. 123-131.

Weigel, D. and E. M. Meyerowitz, (1993). "Activation of floral homeotic genes in *Arabidopsis*," *Science* vol. 261, pp. 1723-1726.

Weinberg, R. A., (1985). "The molecules of life," *Scientific American*, vol. 253.

Weismann, A. (1893). *The Germ Plasm: A Theory of Heredity*, Translated by W. N. Parker and H. Rönnfeldt, London: Walter Scott Ltd.

Welch, W. J., (1993). "How cells respond to stress," *Scientific American*, vol. 268, May.

West-Eberhard, M. J., (1986). "Alternative adaptations, speciation, and phylogeny (A review)," *Proceedings National Academy of Science USA*, vol. 83, pp. 1388-1392.

West-Eberhard, M. J., (1989). "Phenotypic plasticity and the origins of diversity," *Annual Review of Ecological Systems*, vol. 20, pp. 249-278.

Willey, A., (1911). *Convergence in Evolution*, London: John Murray.

Williams, R. W., C. Cavada, and F. Reinoso-Suárez, (1993). "Rapid evolution of the visual system: A cellular assay of the retina and dorsal lateral geniculate nucleus of the Spanish wildcat and the domestic cat," *Journal of Neuroscience*, vol. 13, pp. 208-228.

Williamson, P. G., (1981a). "Palaeontological documentation of speciation in Cenozoic molluscs from Turkana Basin", *Nature*, vol. 293, pp. 437-443.

Williamson, P. G., (1981b). "Morphological stasis and developmental constraint: real problems for neo-Darwinism," *Nature*, vol. 294, pp. 214-215.

Williamson, P. G., (1982). "Punctuationism and Darwinism reconciled? The Lake Turkana mollusc sequence," *Nature*, vol. 296, pp. 608-612.

Williamson, P. G., (1983). "Speciation in molluscs from Turkana Basin," *Nature*, vol. 304, pp. 661-663.

Williamson, P. G., (1985a). "In reply to Fryer, Greenwood and Peake," *Biological Journal of the Linnean Society*, vol. 26, pp. 337-340.

Williamson, P. G., (1985b). "Punctuated equilibrium, morphological stasis and the palaeontological documentation of speciation: a reply to Fryer, Greenwood and Peake's critique of the Turkana Basin mollusc sequence," *Biological Journal of the Linnean Society*, vol. 26, pp. 307-324.

Wilson, A. C., L. R. Maxson, and V. M. Sarich, (1974). "Two types of molecular evolution. Evidences from studies of interspecific hybridization," *Proceedings National Academy of Science USA*, vol. 71, pp. 2843-2847.

Wilson, E. O., 1992. *The Diversity of Life*, Cambridge, Harvard University Press.

REFERENCES

Wu, T. T., E. C. C. Lin, and S. Tanaka, (1968). "Mutants of *Aerobacter aerogenes* capable of utilizing xylitol as a novel carbon," *Journal of Bacteriology*, vol. 96, pp. 447-456.

Wynne-Edwards, V. C., (1965). "Self-regulating systems in populations of animals," *Science*, vol. 147, pp. 1543-1548).

Wynne-Edwards, V. C., (1986). *Evolution Through Group Selection*, London: Blackwell.

Zuker, C. S., (1994). "On the evolution of eyes: Would you like it simple or compound?" *Science*, vol. 265, pp. 742-743.

INDEX

A

accumulative selection • 178.
See also cumulative selection
activation energy • 217–219
activators • 230
adaptation by information loss •
127–128
adaptive • 65, 162, 168, 170,
172, 173
adaptive mutation • 56, 133, 134
adaptive variation • 51, 53, 56,
61, 79, 87, 97, 98, 100, 102–
107, 113, 117, 119, 123, 133–
134, 208
adaptive variation, storage of •
118
Agard, D.A. • 230
Agard, D.A. • *See* Bone, R,
Aharonowitz, Y. • 139
alcohol resistance • 65
alleles • 46, 51, 65
allosteric • 222
amino acids • 31, 35, 36, 45, 96,
97, 116, 117, 137–139, 147,
156, 226–228, 230
ampicillin, resistance • 186
amplification • 69
Antennapedia gene • 236
antibiotic resistance • 138–141,
144, 159. *See also
streptomycin resistance*
Applebury, M.L. • 45

B

Argument from Design • 5–6,
11, 17, 18, 120, 161–162, 174
Ayala, F.J. • 64. *See also* Lenski,
R.E.

back mutation • 72, 73
backbone of DNA molecule •
213, 214
bacteria, mutations in • 149–156
Bahill, A.T. • 129
Baltimore, D. • *See* Darnell, J.E.
Barrande, J. • 130–131
base • 28, 29, 34, 38, 44, 46, 48,
58, 63, 81, 213–216, 223, 224,
226–228
base pairs • 29, 38, 46
bats, evolution of • 84, 111
Baylor, D.A. • 129
Beacham, T.D. • 208
Beeman, R.W. • 143
Benson, S.A. • 159, 190
Benveniste, R. • *See* Davies, J.
Bergström, S • *See* Normark, S.
Berson, E.L. • *See* Dryja, T.P.
beta-galactosidase • 158
Bishop, J.A. • 67
bit • 72–74, 82– 83, 215
Bithorax gene • 236
Blind Watchmaker • 161–174
Bone, R. • 155, 156
Bonner, J.T. • 128
Boyajian, G. • 131

Darwinian assumptions • 104–
105, 107, 113, 117, 118, 160,
163, 169–170
dauer modification • 243
Davies, J. • 138, 139
Dawkins, R. • ix, x, 61, 96, 101,
113, 121, 161–174
de Candolle, Pyrame • 8
De Robertis, E.M. • 237
de Vries, Hugo • 19
Delage, Yves • 175
Delbrück, Max • 189
deletion • 63, 69, 70, 186, 188,
189
deoxyribonucleic acid • *See*
DNA
descent, theory of • 11, 18, 20,
21, 77, 109, 110, 115, 128
development • 26, 178, 180, 183,
193, 195, 196, 233–238
Dickens, Charles • 59
differentiation • 26, 193, 195,
235, 236
DNA • 27–31, 36, 48, 57, 58,
62–64, 69, 70, 77, 81–83, 88,
89, 92, 97, 113, 116–118, 122,
139, 141, 152, 163, 168, 181,
183–188, 192, 193, 195, 198,
199, 201, 202, 213–216, 223–
227, 230, 233, 235, 237
DNA replication • 38–40
DNA structure • 213–215
Dobzhansky, T. • 20, 56, 101,
105, 185
dogma of the NDT • 51
Donelson, J.E. • 28
Drake, J.W. • 92
Drosophila • 27, 28, 35, 65, 67,
105, 185, 235–237
Drummond, H. • 211, 212
Dryja,T.P. • 45
duplications • 186

E

E. coli • 187, 190, 191
Edgar, B.A. • *See* O'Farrell, P.H.
Edlund, T. • *See* Normark, S.
Eldredge, N. • 68
electricity in fish, evolution of •
111
Eliot, C.W. • 14
embryo • *See development*
Endler, J.A. • *See* Reznick, D.A.
Engelberg, J. • 129
environmental cue • x, 182, 183,
192, 194, 199, 201, 208, 240,
241, 243, 244
enzyme • 32–34, 36, 45, 61, 88,
89, 92, 107, 114, 115, 118,
137–139, 149–151, 153, 154,
156–158, 182, 187–189, 191–
193, 216–220, 233
Euglena • 111
eukaryotes • 225
evolution of complexity • 61,
72–75, 83, 112, 113, 120, 128–
131, 133, 164
evolution of information • 70–
72. *See also information,*
evolutionary buildup
evolution of information • 89
exons • 225
exponential growth • 52–53
eye, evolution of • 60, 111–112,
129

F

feedback inhibition • 32, 221,
232
Fersht, A.R. • 92. *See also*
Grosse, F.
finches • 202–205
Fischbach, G.D. • 112
fish & cephalopod comparison •
207

INDEX

product • 31
prokaryotes • 225
promoter • 185, 223, 230, 233
protein • 31–34, 66, 69, 92, 96,
  97, 107, 122, 134, 137, 143,
  144, 159, 185, 216, 220, 222–
  228, 230–239, 240
protein synthesis • 35–36, 139,
  141, 144–146, 228–230
punctuated equilibrium • 68

Q

quantitative traits • 134, 144–
  148
Quiring, R. • 112, 114

R

Radicella, J.P. • 191
random • 87–91, 94, 98, 107,
  108, 110, 113, 119–121
random events • 50, 76
random variation • vii, x, xi, 50,
  51, 58, 65, 67, 75, 88, 90, 121,
  160, 175–177, 179, 180, 182,
  189, 190, 197, 203, 204
Raven, C.E. • 6
Ray, John • 4
RDH • *See ribitol
  dehydrogenase*
reaction energy • 217
rearrangements • *See genetic
  rearrangements*
recombination • 40, 51, 62, 64,
  65, 118, 119, 185, 186
redundant code • 226
regulatory genes • 35, 45, 69, 70,
  72, 88, 90, 146, 148, 153, 180,
  188, 231
regulatory protein • 37, 45, 145,
  146, 147, 153, 157, 193, 231,
  235, 240, 241
Reichel, E. • *See* Dryja, T.P.

Reinoso-Suárez, F. • *See*
  Williams, R.W.
Rensch, B. • 179
repressors • 230
retroviral-like element • 88–89
retrovirus • 88
Reznick, D.A. • 205, 206
ribitol • 149–154, 157
ribitol dehydrogenase • 150–153
ribonucleic acid • *See RNA*
ribosome • 36, 139, 141, 144,
  228
Richter, C.P. • 195
Rigby, P.W.J. • 149. *See also*
  Burleigh, B.D.
RLE • *See retroviral-like
  element*
RNA • 34–35, 88, 92
RNA polymerase • 36, 45, 223,
  224, 230
RNA transcription • 223–226
Robertson, M. • 237
rodent jaw evolution • 195–196
Rosen, D.E. • 182
Rosenberg, S.M. • 191
Rosenberger, R.F. • 159
Roth, R.R. • 191
Rowland, M.W. • 144
Ruse, M. • 63
Rushton, W.A.H. • 129
Rust, J.W. • *See* Hermas, S.A.

S

Saint-Hilaire, Geoffry • 7
salicin metabolism, evolution of
  • 188–189
Salmonella • 185
Sandberg, M.A. • *See* Dryja,
  T.P.
Sarich, V.M. • *See* Wilson, A.C.
Saunders, P.T. • 121, 176, 182,
  203